Decision Making, Planning, and Control Strategies for Intelligent Vehicles

Synthesis Lectures on Advances in Automotive Technology

Editor
Amir Khajepour, *University of Waterloo*

The automotive industry has entered a transformational period that will see an unprecedented evolution in the technological capabilities of vehicles. Significant advances in new manufacturing techniques, low-cost sensors, high processing power, and ubiquitous real-time access to information mean that vehicles are rapidly changing and growing in complexity. These new technologies—including the inevitable evolution toward autonomous vehicles—will ultimately deliver substantial benefits to drivers, passengers, and the environment. Synthesis Lectures on Advances in Automotive Technology Series is intended to introduce such new transformational technologies in the automotive industry to its readers.

Decision Making, Planning, and Control Strategies for Intelligent Vehicles
Haotian Cao, Mingjun Li, Song Zhao, and Xiaolin Song
2020

Autonomous Vehicles and the Law: How Each Field is Shaping the Other
Ayse Buke Hiziroglu
2020

Cyber-Physical Vehicle Systems: Methodology and Applications
Chen Lv, Yang Xing, Junzhi Zhang, and Dongpu Cao
2020

Reinforcement Learning-Enabled Intelligent Energy Management for Hybrid Electric Vehicles
Teng Liu
2019

Deep Learning for Autonomous Vehicle Control: Algorithms, State-of-the-Art, and Future Prospects
Sampo Kuuti, Saber Fallah, Richard Bowden, and Phil Barber
2019

Decision Making, Planning, and Control Strategies for Intelligent Vehicles

Haotian Cao, Mingjun Li, Song Zhao, and Xiaolin Song

ISBN: 978-3-031-00378-3   paperback
ISBN: 978-3-031-01506-9   ebook
ISBN: 978-3-031-00010-2   hardcover

DOI 10.1007/978-3-031-01506-9

A Publication in the Springer series
SYNTHESIS LECTURES ON ADVANCES IN AUTOMOTIVE TECHNOLOGY

Lecture #12
Series Editor: Amir Khajepour, *University of Waterloo*
Series ISSN
Print 2576-8107   Electronic 2576-8131

Decision Making, Planning, and Control Strategies for Intelligent Vehicles

Haotian Cao
Hunan University, China

Mingjun Li
Hunan University, China

Song Zhao
University of Waterloo, Canada

Xiaolin Song
Hunan University, China

SYNTHESIS LECTURES ON ADVANCES IN AUTOMOTIVE TECHNOLOGY #12

ABSTRACT

The intelligent vehicle will play a crucial and essential role in the development of the future intelligent transportation system, which is developing toward the connected driving environment, ultimate driving safety, and comforts, as well as green efficiency. While the decision making, planning, and control are extremely vital components of the intelligent vehicle, these modules act as a bridge, connecting the subsystem of the environmental perception and the bottom-level control execution of the vehicle as well. This short book covers various strategies of designing the decision making, trajectory planning, and tracking control, as well as share driving, of the human-automation to adapt to different levels of the automated driving system.

More specifically, we introduce an end-to-end decision-making module based on the deep Q-learning, and improved path-planning methods based on artificial potentials and elastic bands which are designed for obstacle avoidance. Then, the optimal method based on the convex optimization and the natural cubic spline is presented.

As for the speed planning, planning methods based on the multi-object optimization and high-order polynomials, and a method with convex optimization and natural cubic splines, are proposed for the non-vehicle-following scenario (e.g., free driving, lane change, obstacle avoidance), while the planning method based on vehicle-following kinematics and the model predictive control (MPC) is adopted for the car-following scenario. We introduce two robust tracking methods for the trajectory following. The first one, based on nonlinear vehicle longitudinal or path-preview dynamic systems, utilizes the adaptive sliding mode control (SMC) law which can compensate for uncertainties to follow the speed or path profiles. The second one is based on the five-degrees-of-freedom nonlinear vehicle dynamical system that utilizes the linearized time-varying MPC to track the speed and path profile simultaneously.

Toward human-automation cooperative driving systems, we introduce two control strategies to address the control authority and conflict management problems between the human driver and the automated driving systems. Driving safety field and game theory are utilized to propose a game-based strategy, which is used to deal with path conflicts during obstacle avoidance. Driver's driving intention, situation assessment, and performance index are employed for the development of the fuzzy-based strategy.

Multiple case studies and demos are included in each chapter to show the effectiveness of the proposed approach. We sincerely hope the contents of this short book provide certain theoretical guidance and technical supports for the development of intelligent vehicle technology.

KEYWORDS

intelligent vehicle, decision making, path planning, speed planning, robust trajectory tracking control, driving intention, human-automation cooperative driving, deep Q-learning, driving hazard potential, convex optimization, dynamical game theory

Contents

Acknowledgments

This work was supported by the National Natural Science Foundation of China (grant numbers 51905161 and 51975194). We would like to thank Mr. Xin Sheng and Mr. Zhiqiang Chen for their kind help while writing this short book. We are also thankful to Morgan & Claypool Publishers for providing the opportunity and support for this book.

Haotian Cao, Mingjun Li, Song Zhao, and Xiaolin Song
June 2020

CHAPTER 1

Introduction

With the increasing numbers of vehicles globally, and the resulting problems such as traffic jams, traffic accident, and environmental pollution are becoming more and more serious. According to data from the World Health Organization (WHO), road traffic crashes result in the deaths of approximately 1.35 million people around the world each year and leaves between 20 and 50 million people with non-fatal injuries [1]. The WHO predicts that road traffic injuries will become the seventh leading cause of death in the world by 2030 [2]. As for the circumstance of public transport safety in China, despite a significant decline in road traffic accidents in recent years, the overall traffic casualties are still high. There were over 24,000 road traffic accidents, causing more than 63,000 deaths in 2018 [3]. While the situation in the United States is much less severe, the number of traffic fatalities in 2016 was only 22,000 with 220 million passenger cars [4]. Therefore, concerns about transport safety have been raised and are a focus of governments, industries, and academia worldwide.

A variety of solutions to the Intelligent Transportation Systems (ITS) and intelligent vehicles have been proposed and studied to solve the hidden invisible dangers caused by traffic accidents, and the intelligent vehicle is considered one of the most promising ways to ultimately reduce the possibility of traffic accidents. Thus, it has become the research frontier of the ITS nowadays. Generally, the intelligent vehicle is a complicated system which integrates the subsystem of environment perception, information fusion, decision making, trajectory planning, and control. The system is expected to assist (e.g., Society of Automative Engineers (SAE) automatic driving levels 1-3) or even replace the human driver (e.g., SAE automatic driving levels 4-5) and to complete various driving tasks, such that it could improve the efficiency and driving safety in different driving situations, and effectively alleviate the situation of traffic congestion in urban areas. We will mainly focus on the algorithms on the decision making, planning, and control for intelligent vehicles on structured roads (e.g., high ways, urban roads) in this short book, which are vital functional modules for the intelligent vehicle, and act as a bridge between the environment perception, sensor fusion, and vehicle lower executions. We hope the contents of this short book can provide certain theoretical guidance and technical support for the development of intelligent vehicle technology.

1.1   BRIEF INTRODUCTIONS ON TRAJECTORY PLANNING FOR INTELLIGENT VEHICLES

Generally speaking, the trajectory information contains the profiles of both the path and the speed. Intuitively, the trajectory optimization can be realized by solving the following optimization problem while minimizing the cost function J, which has an analytic form as,

$$J = \int_0^{t_f} l(\boldsymbol{x}(t), \boldsymbol{u}(t), t)dt + V(\boldsymbol{x}(t_f), t_f), \tag{1.1}$$

where t_f denotes the terminal time and $l(\cdot)$ denotes the cost function related to the trajectory planning (e.g., path length, curvatures, control amplitudes), which is generally a function of the states $\boldsymbol{x}$, the control $\boldsymbol{u}$, and the time t, while $V(\cdot)$ denotes the terminal costs at the endpoints. Moreover, the states $\boldsymbol{x}$ satisfy the following dynamical system:

$$\dot{\boldsymbol{x}}(t) = \boldsymbol{f}(\boldsymbol{x}(t), \boldsymbol{u}(t)), \tag{1.2}$$

where $\boldsymbol{f}$ denotes the mapping function of the dynamical system (e.g., the vehicle); besides, there usually exists extra equality constraints (e.g., initial states, terminal states) and inequality constraints for the states and controls, which are defined by

$$\begin{cases} \boldsymbol{g}(\boldsymbol{x}(t), t) = 0 \\ \boldsymbol{h}(\boldsymbol{x}(t), t) > 0. \end{cases} \tag{1.3}$$

Combining (1.1) and (1.3), the trajectory optimization problem can be summarized as

$$\min_{\boldsymbol{u}} J, \quad \text{s.t. } (1.2), (1.3). \tag{1.4}$$

In other words, it means finding the optimal control sequence $\boldsymbol{u}$ which minimizes the cost function J within the time range $[0, t_f]$ while subjected to related constraints (1.2)–(1.3). Bellman proposed a global optimization method that discretizes (1.4) that can greatly reduce the computational consumption requirements in the 1950s, by the so-called dynamic programming method based on Bellman's Principle of Optimality [5]. This principle breaks a dynamic optimization problem into a sequence of simpler subproblems, thereby significantly reducing the difficulty of the mathematical processing. However, a direct or indirect solution for (1.4) is not an easy task. General approaches for solving this optimal control problem include the calculus of variations [6], single-shooting or multiple shooting [7], the dynamic programming [8, 9], etc. Still, if you need to take into account that a vehicle is a complex dynamical system, the time-consuming task of finding a solution can be even longer, even in numerical methods. That is why a direct optimization as expressed in (1.4) is not that popular in real applications. The trajectory optimization for controlling a complex vehicle system is then decomposed into the path (speed) planning and the path (speed) following, which will be elaborated upon in the following contents.

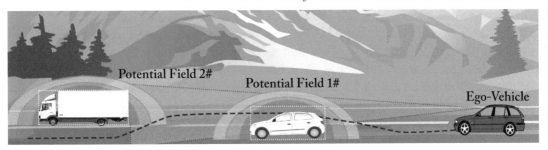

Figure 1.1: Concept of path planning by the artificial potential.

At present, most of the path-planning methods are inherited from the robotic field. There have been many planning methods, such as grid-based methods, discrete optimization approaches, geometric planning methods, potential-field approaches, heuristic search algorithms, intelligent swarm algorithms, and randomized path planning such as a probabilistic road map. For example, path-planning methods from teams participating in the DARPA Urban Challenge [10] include, but are not limited to the A*/D* algorithm, RRT, and artificial potential. As for the speed planning for intelligent vehicles, planning algorithms might be different due to different driving situations. For the non-following scenario where the host vehicle usually moves free within the road boundaries and overtakes the preceding vehicle or obstacles if present, most applications will adopt triangular or trapezoidal velocity profiles [10, 11]. However, in the case of intelligent vehicle's following, the algorithm of speed planning is more like an upper level of the Adaptive Cruise Control (ACC). As many methods can be realized for adaptive following, readers may refer to [12–15] for more details. So as to provide a simple introduction in this book, we will primarily discuss the several most well-known and widely applied path-planning algorithms.

The Virtual Potential Method

The virtual potential method for path planning was originally proposed by Khatib [16] for robot applications. As shown in Figure 1.1, the movement of the vehicle is governed by virtual force potentials, where the obstacles produce a repulsive effect on the ego-vehicle, while the target point generates an attractive effect on the ego-vehicle. As a result, the ego-vehicle will move to the target point while attempting to collide with the obstacles. Moreover, the attractive effect and the repulsive potential will increase with the decrease of their distance gap. The repulsive potential generally reaches maximum value at the boundary of the obstacles.

One of the biggest advantages of the virtual potential method for path planning is that it does have excellent real-time performances; with simple structures and convenient bottom controls, it is widely applied for the obstacle avoidance in historical times. However, it also has some shortcomings, e.g., it suffers from the problem of "local minima trap." The range of the potential field is relatively large, while the repulsive potentials are only partial. When the distance between the main vehicle and the obstacle exceeds the obstacle's influence range, at this time, the vehicle is not easily affected by the repulsive potentials. Therefore, the potential field method is prone to fall into the local minima due to a lack of global information, in that case, resulting in oscillations to the planned path. Besides, the more obstacles there are, the greater the possibility of generating multiple local minima, resulting in increased planning difficulty and even total failure. To overcome this disadvantage, Brandt et al. focused on lateral guidance strategies based on the elastic band and the force potentials [17, 18]. Cao et al. present a harmonic velocity potential-based method for active obstacle avoidance [19]. Li et al. use the elastic band to generate a feasible path for shared driving [20]. Recently, Khajepour et al. present a 3D virtual dangerous potential field constructed as a superposition of trigonometric functions of the road and the exponential function of obstacles [21, 22].

1.1.1   THE RAPIDLY EXPLORING RANDOM TREE (RRT) METHOD

Rapidly exploring random tree (RRT) is a path-planning algorithm for searching non-convex high-dimensional space efficiently by constructing space-filling trees randomly. The searching tree is constructed step-by-step from randomly drawn samples in the search space; in essence, it tends to grow toward most undetected areas. The RRT algorithms, originally developed by LaValle [23, 24], are widely used in the areas of robotic motion planning because they can handle obstacles and differential constraints (nonholonomic dynamics) very easily. It can be regarded as a technique for generating open-loop trajectories in a nonlinear system with state constraints. Moreover, a new RRT-connected by LaValle and Kuffner is further proposed, where two random trees grow from the initial and terminal states at the same time, and one of the trees expands and attempts to connect the nearest node of the other tree to expand the new node during each iteration. They then exchange the order and repeat the previous process. Such that the RRT-connected algorithm has advantages in the searching speed and the efficiency compared with the original one, especially when the starting and terminal states located in constrained areas, with such a heuristics strategy, these two searching trees can escape from their respective restricted areas by rapidly expanding toward each other.

Most of the RRT algorithms have been modified and improved for better adaptations in the vehicle path planning. The RRT algorithm has derived many variants until now, e.g., Feng Laichun et al. proposed an RRT algorithm based on the shortest path generated by the A* to solve the problems of low search efficiency and unreasonable nearest neighbor search function in the original RRT [26]. Ma et al. proposed an improved RRT [27] for the urban environment to overcome the shortcomings of the generated tortuous path. Based on different driving behaviors such as straight, left turn, right turn, etc., a pre-built template is used to initialize the tree as guidance information and improve search efficiency simultaneously. Song et al. generated random sampling points obeying the Gaussian distribution according to the expected path, then introduced a heuristic search mechanism in the original RRT to find a suitable obstacle avoidance path. Moreover, the B-spline curve interpolation method is used for trajectory smoothing, thereby the improved RRT algorithm is more in line with vehicle obstacle avoidance at high speeds [28]. Yuan et al. proposed a motion planning and obstacle avoidance of industrial manipulators by the RRT-connected algorithm [29].

1.1.2   INTELLIGENT SWARM ALGORITHM

Intelligent swarm algorithms are an emerging evolutionary computing technology, which has raised increased attention from many researchers. We will introduce several widely used and representative algorithms for vehicle path planning, including ant colony optimization, artificial fish swarm, and intelligent water drops.

(1) **The Ant Colony Optimization Algorithm:** The ant colony optimization (ACO) algorithm was first proposed by Dorigo [30] and its basic idea comes from the shortest path principle of ants foraging in nature. Ants can release a kind of biological hormone called a pheromone to exchange foraging information on the path they traverse when looking for food sources so that other ants within a certain range can detect it. When more and more ants pass on certain paths, there are more and more pheromones accumulated, and the probability that the ants choose this path is higher. Those behaviors emerge as a form of autocatalysis. Surprisingly, the ant does not have to find the shortest path for a single ant, but only chooses according to probability; for the entire ant colony system, they can achieve the objective effect of finding the optimal path so that is what is called swarm intelligence. Generally, we suppose the whole environment in which ants are located in a virtual world, including obstacles, other ants, and pheromones. The ant perception range is usually set to the grid world, and each individual can only perceive the environmental information within its certain ranges. Pheromones can be divided into the food pheromone left by the ants who found the food and the pheromone left by the ants who found the nest. In adddition, the pheromone will be lost at a certain rate in the environment. Thus, it is suitable for the optimization in path-planning problems due to the above mechanism, and it is favored by many scholars. For example, Fan et al. used it for robotic path planning in the complex working environment by adjusting the relative importance of pheromone and heuristic information [31]. Wu et al. designed the goal of minimum driving time for the off-road path planning problem of vehicles by the ant colony algorithm with multi-strategies [32]. Xiao et al. proposed a terrain model with uniform grids that is subjected to the Unmanned Aerial Vehicle (UAV) constraints, and designed a comprehensive performance evaluation system, which includes the cost of path deviations, altitudes, terrain avoidances, threats, and safeties, then uses the ant colony algorithm to improve the efficiency of the path planning to reach the destination in a shorter time [33]. Jabbarpour et al. proposed a path-planning strategy using the ant colony algorithm for optimal energy saving [34] for saving energy consumption of unmanned

ground vehicles. Moreover, the literature [35, 36] also focuses on the vehicle path planning by the ant colony optimization, which will not be elaborated on here.

(2) **The Artificial Fish Shoals Algorithm:** The artificial fish shoals (AFS) algorithm is a new type of intelligent swarm optimization technique motivated by intelligent behaviors of fish shoals [37]. The algorithm is based on the fact that the water area with the most nutrients has the largest number of individual fish survivals. Therefore, this feature is used to simulate the foraging behavior of fish shoals to achieve an optimal search. Basic behaviors of fish shoals include the foraging, clustering, and rear-ending. The optimization process starts from the basic behavior modeling of individual, fining the local optimizations, and, finally, obtain the optimal global result. The algorithm has a bulletin board to record the optimal state of each fish, and the state of each fish will be compared with the value recorded in the bulletin board at every iteration. Then, it will update the state in the bulletin board with current states if it is better, otherwise, the state of the bulletin board remains unchanged. Therefore, the value of the bulletin board is the optimal solution when the iteration of the entire algorithm ends. The criterion for the algorithm to terminate the iteration is when the obtained mean square error for multiple consecutive times is less than the predefined allowable error, or the number of fish gathered in a certain area reaches a certain ratio or the average value obtained for multiple consecutive times does not exceed the sought value or reaching the maximum number of iterations. Eventually, the optimal record of the bulletin board is the output, otherwise, the iteration continues. Currently, the AFS algorithm can be used for solving the robotic path-planning problems such as the mine rescue robotics [38] and ship collision avoidance [39]. Zhang et al. introduced an improved AFS for robotic path-planning, where the direction operator improves the accuracy and success rate of the behaviors of the fish shoals of foraging, grouping, and rear-end, the immune memory operations are added to improve the algorithm's global search capability and reduce the probability of local extrema [40].

(3) **The Intelligent Water Drop Algorithm:** Intelligent Water Drop (IWD) is a new swarm intelligence algorithm proposed by Hosseini [41]. In nature, the scouring effect of water flow will form a gully on the ground that can bypass obstacles and successfully reach a low-terrain location. The water flow can be regarded as a group composed of unit water droplets, and each water droplet has attributes of a variable velocity and certain sediment. When water droplets flow through one area, the water droplets are more likely to pass through a river bed with less sediment. In this way, water droplets will gain larger increments in speed and wash away more sediment, eventually forming a feedback mechanism for the path of other droplets. Therefore, the key idea of the IWD is to simulate the interaction of the water flow and sediment to form a channel, such that solving complex problems in the field of computational science. The algorithm has been successfully applied to the Traveling Salesman Problem (TSP) [41], Multiple

Knapsack Problem (MKP) problems [42], and Vehicle Routing Problem (VRP) [43]. The IWD can be also applied for the application of path planning. For example, Duan et al. utilized the improved IWD for the path planning of the UAV to avoid dangerous threats [44, 45]. Similarly, to add sufficient heuristics to find the optimal path for obstacle avoidance, Song et al. proposed an improved IWD in the mechanisms of selection and updates [46].

1.2   BRIEF INTRODUCTIONS ON TRAJECTORY FOLLOWING CONTROL

There are many types of closed-loop trajectory tracking applications used for the vehicle. In general, conventional methods of trajectory following can be roughly divided into the compensation models based on the classical transfer function, single-point preview-following strategy, multi-point preview model, and intelligent control models (e.g., the fuzzy, neural networks). Many models have been integrated into some commercial multibody dynamics software such as Adams®, CarSim®, LMS ImagineLab®, etc. We will introduce several most well-known ones in the following contents.

1.2.1   CLASSICAL TRANSFER FUNCTION MODEL

From the cybernetic point of view, the formulation of the vehicle's steering strategy can be understood as the driver's attempt to eliminate the lateral deviation error Δy_p at the preview distance L, as shown in Figure 1.2. In terms of transfer functions, it can be shown in a more general form by the control loop, where $P(s)$ represents the preview strategy of the driver for the desired path, $H(s)$ the control characteristics of the driver, $G(s)$ the transfer function(s) of the vehicle system dynamics, and $B(s)$ denotes the feedback element. McRuer et al. proposed a compensation tracking model for their working company, Systems Technology Inc. in 1965 [47–49] (that is why it is called the STI model), which derived from the research of the pilot-aircraft closed-loop system and incorporates the driver's reaction time. The driver's characteristic function can be described by (1.5):

$$H(s) = \frac{K(T_L s + 1)}{(T_I s + 1)(T_N s + 1)} e^{-\tau s},\tag{1.5}$$

where τ denotes the driver's reaction time, T_N denotes the lag of driver's neuromuscular system, and T_L, T_I, and K denote designed parameters related to the driver control system. The STI model assumes that the driver performs compensations and corrections based on the lateral deviation between the current vehicle trajectory and the expected one, so as, to study the driver's behavior characteristics.

In addition the well-known driver crossover model [50] provides an effective way to describe the driver's lateral control tasks around the crossover frequency, which is usually applied

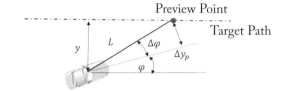

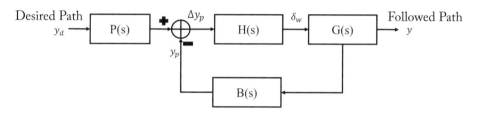

Figure 1.2: Control scheme of the preview-following model.

to the parameter recognition of the driver's driving behavior. More specifically, it takes into account the combined dynamics of the human driver and the vehicle system, using the following form around the crossover frequency:

$$H(s)G(s) = \frac{\omega_c}{s} e^{-\tau s}, \tag{1.6}$$

where ω_c denotes the crossover frequency, and τ denotes the lag of the system (transport delay time caused by the driver neuromuscular system). That indicates an integral behavior and an adapted phase shift due to the dead time element near the crossover frequency.

1.2.2   SINGLE-POINT PREVIEW-FOLLOWING MODEL

Considering that the driver can usually only observe the path in a limited range, most driver models only optimize the path error within the limited preview distance. The tracking error can be evaluated by multiple indexes, such as the lateral offset error, the path heading error, etc. One of the most well-known preview-following models is the MacAdam's "single-point preview" strategy based on the optimal control theory [51, 52], which has been integrated with a well-known vehicle dynamics simulation software CarSim®. Let us take the simplest two degrees-of-freedom (2DOF) vehicle model with a constant moving speed as an example, where the vehicle dynamic system can be described by a linear state space,

$$\begin{cases} \dot{X} = \mathbf{A_v} X + \mathbf{B_v} \delta_w \\ y = \mathbf{C_v} X, \end{cases} \tag{1.7}$$

where X denotes the state variable of the 2DOF vehicle model, δ_w denotes the wheel steering angle, which is regarded as the control input of the system, and y denotes the system output, which in this case, the lateral offset of the vehicle. $\mathbf{A_v}, \mathbf{B_v}, \mathbf{C_v}$ denotes the 2DOF vehicle system matrix. For any t within the preview time T_p, the response of the vehicle's lateral offset with zero states is calculated by

$$y(t + \theta) = f(\theta)X(t) + g(\theta)\delta_w(t) \tag{1.8}$$

$$f(\theta) = \mathbf{C_v}e^{\mathbf{A_v}\theta}, g(\theta) = \mathbf{C_v}\left(\int_0^t e^{\mathbf{A_v}\theta}\mathbf{B_v}\right). \tag{1.9}$$

Consider one objective function related to the tracking error within the preview time,

$$J = \frac{1}{T_p}\int_t^{t+T_p}(y_d(\theta) - y(\theta))^2\delta(T_p)d\theta, \tag{1.10}$$

where y_d represents the expected lateral offset, and $\delta(T_p)$ represents the Dirac function evaluated within $[0, \delta(T_p)]$, by solving the equation $\frac{\partial J}{\partial u} = 0$, and will obtain the optimal steering angle δ_w^*,

$$\delta_w^* = \frac{y_d(t + T_p) - f(T_p)X(t)}{g(T_p)}. \tag{1.11}$$

Another famous single-point preview-following model is the optimal curvature model proposed by Guo [53, 54], which believes the transfer function (1.12) will be satisfied at low frequencies,

$$\frac{Y(s)}{Y_d(s)} = P(s) \cdot F(s) \approx 1, \tag{1.12}$$

where $P(s) = e^{(T_p s)}$ is called the predictor, $F(s)$ is the following controller, $Y(s)$ denotes the Laplace transform of the actual lateral offset, and $Y_d(s)$ denotes the Laplace transform of the ideal lateral offset. The model establishes relationships among the vehicle-handling characteristic parameters, the driver characteristic parameters, and the path-following model parameters. Experimental results confirm that the whole tacking system achieves better following results only when $P(s) \cdot F(s) \approx 1$ is satisfied in the low-frequency domain. The flow process of the single-point preview-following the model proposed by Guo is shown in Figure 1.3, which takes into account the driver's physiological constraints and characteristics of vehicle dynamics [55]. The driver's physiological limitation mainly comes from the driver's reaction lag, which includes the driver's neural response lag and the operation lag. The former is usually a pure lag that describes the driver's perception of various information, and it can be represented by the transfer function $e^{-T_d s}$, where T_d denotes the neural response lag time. While the driver's operation lag can usually be described by a first-order inertial element $1/(1 + T_h s)$, where T_h denotes the action response lag time. Besides, considering that the driver can adjust the steering input according to the change of the path, which means the driver can perform simple differential calculations, such that the differential loop $C_d(1 + T_c s)$ of correction is added in the model, where C_d, T_c denote the designed parameters.

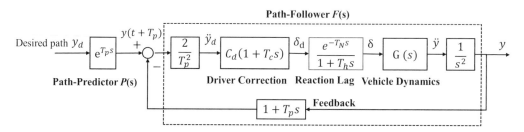

Figure 1.3: The processing flow of the single-point preview-following model by Guo.

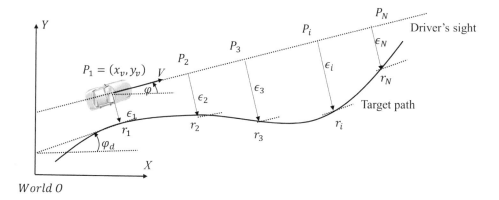

Figure 1.4: The concept of the multiple-point preview.

1.2.3   MULTI-POINT PREVIEW-FOLLOWING MODEL

The very earliest complication of the multi-point preview following was first proposed by Sharp [56, 57]. As shown in Figure 1.4, it is assumed that the driver will process multiple N-preview points ahead simultaneously within the preview distance $L = V T_p$ while following the target path. Moreover, each preview point P_i corresponds to a lateral offset ϵ_i from the driver's sight path to the target path. Sharp believes that the steering wheel angle is a linear combination of the current lateral offset of the vehicle, the difference between the current yaw angle of the vehicle and the corresponding heading of the target path ($\varphi - \varphi_d$), and other lateral offsets ϵ_i, namely,

$$\delta_w = K_0(\varphi - \varphi_d) + \left(K_1\epsilon_i + \sum_{i=2}^{N} K_i\epsilon_i\right), \tag{1.13}$$

where $K_i(i = 0, 1, 2, \ldots, N - 1)$ denote the control gain, usually the farther the position of the preview point, the smaller the corresponding control gain, which will be determined by the Linear Quadratic Regulator (LQR) method introduced later.

The LQR theory is the earliest and one of the most mature closed-loop optimal control methods, which obtains the optimal control law of state linear feedback easily by solving a Riccati equation. As the vehicle moves forward, the road preview system will read a new preview point ϵ_{N+1} at the next instant, while the first preview point data at the current time will be moved out. Such that the road preview system would be described by a shift register, namely,

$$Y(k + 1) = D_r Y(k) + E_r w(k), \tag{1.14}$$

where

$$D_r = \overbrace{\begin{bmatrix} 0 & 1 & 0 & \cdots & 0 \\ 0 & 0 & 1 & \cdots & 0 \\ \vdots & \vdots & \vdots & \ddots & \vdots \\ 0 & 0 & 0 & \cdots & 1 \\ 0 & 0 & 0 & \cdots & 0 \end{bmatrix}}^{N+1} \quad E_r = \begin{bmatrix} 0 \\ 0 \\ \vdots \\ 0 \\ 1 \end{bmatrix} \Bigg\} (N + 1),$$

where $Y = [\epsilon_1, \epsilon_2, \ldots, \epsilon_i, \ldots, \epsilon_N]^T$ denotes the preview states and $w(k)$ denotes the new lateral offset at the next instant ϵ_{N+1}, while D_r, E_r denote the system matrix.

Besides, the 2DOF vehicle dynamics system (1.8) can be discretized into a linear state-space form,

$$X(k + 1) = A_d X(k) + B_d \delta_w(k). \tag{1.15}$$

Combining with the road preview model (1.14), the final state-space form of the road-vehicle system is written by

$$\begin{bmatrix} X(k + 1) \\ Y(k + 1) \end{bmatrix} = \begin{bmatrix} A_d & 0 \\ 0 & D_r \end{bmatrix} \begin{bmatrix} X(k) \\ Y(k) \end{bmatrix} + \begin{bmatrix} B_d \\ 0 \end{bmatrix} u(k) + \begin{bmatrix} 0 \\ E_r \end{bmatrix} w(k) \tag{1.16}$$

Let $x(k) = \begin{bmatrix} X(k) \\ Y(k) \end{bmatrix}$, such that

$$x(k + 1) = A x(k) + B \delta_w(k) + E w(k). \tag{1.17}$$

In addition, we treat the tracking error of the preview points as the system outputs z, such that

$$\begin{cases} x(k + 1) = A x(k) + B \delta_w(k) + E w(k) \\ z(k) = C x(k). \end{cases} \tag{1.18}$$

If the pair (A, B) can be stabilized, and (A, C) can be observable, then there exists an optimal state feedback control rule u^*, such that,

$$u^* = -K x(k), \, K = (R + B^T P B)^{-1} B^T P A, \tag{1.19}$$

where P satisfies the following discrete Riccati equation;

$$P = A^T P A - A^T P B (R + B^T P B)^{-1} B^T P A + Q \qquad (1.20)$$

with the optimal control law u^*. The following cost function related to the tracking errors and the control amplitude can be minimized, namely,

$$J = \lim_{n \to \infty} \sum_{k=0}^{n} \left(z^T(k) Q z(k) + u^T(k) R u(k) \right), \qquad (1.21)$$

where Q, R denote the weight matrix for the tracking error and the steering-wheel-angle respectively.

However, the LQR method has certain limitations that it cannot effectively handle the system with constraints. Therefore, the most popular path-following model is based on the model predictive control (MPC) method [58–63] in recent years. We will discuss in detail the trajectory following control based on the MPC in later chapters. There are also some other control methodologies can be also used for the path-tracking, such as the fuzzy method [64], neural networks [65], H-∞ control [66], etc. Similarly, most of the advanced control methods, such as the LQR and MPC, could be also used for velocity tracking, however, we do not intend to discuss the details here. Nowadays, the design of a personalized automatic driving system or assisted driving system based on driving styles is valued more and more. For example, Bifulco et al. [67, 68] believe that the longitudinal acceleration decided by the driver can be a linear function of the following distance, relative speed, and the ego-vehicle speed, which indicates the driving style of different drivers can be modeled accordingly. A personalized adaptive cruise (ACC) system is proposed which is based on the evaluation and identification of personal driving styles [69]. Xu et al. [70] established a speed control model with consideration of the driving style based on vehicle test data. Zhu et al. [71] proposed a personalized vehicle lane change assistance system that integrates the driver behavior recognition strategies.

1.3   HUMAN-AUTOMATION COOPERATIVE DRIVING

Nowadays, the highly automated driving systems have various limitations when applied into commerce, such as technological difficulties [72], and social dilemmas [73]. In recent years, human-automation cooperative driving systems have been received growing attention since these systems provide possibilities that human drivers and automated driving systems can control the vehicle together to realize the same tasks and take advantages of both two control participants [74, 75]. In human-automation cooperative driving systems, the human driver and automated driving system have continuous interaction [76], thus the control authority and conflict management between two control participants are two main problems to be discussed.

The prior work about control authority pay attention on the design of certain rules and model-driven shared control methods that are developed based on the assessed driving risk from environment information or detected driving behaviors from the human driver. Nguyen et al. propose an U-shape function considering the drivers' driving workload and performance to calculate the needed control signal for lane-keeping assistance [77]. The research in [20, 78] design cooperative coefficient rules that are used to allocate the control authority between human driver and path-following controller according to the evaluated driving risk and driving intention. As for driving behaviors in human-automation cooperative driving control systems, a MPC-integrated method is proposed to combine the driving control action with other optimized objectives, such as stability performance [79] and lane departure risk [80].

In terms of human-automation conflict management, driver-vehicle model and game theory are widely used to model the human-automation interaction. Nguyen et al. agree that the driver-vehicle model can improve the mutual understanding of the human-automation interaction [81]. Sentouh et al. integrate a human-in-the-loop model into the human-automation cooperative control framework to construct the complete driver-vehicle model and reduce control conflicts [82]. In addition, game theory has been widely utilized to address human-automation conflicts since it is an effective approach to deal with decision-making problems with two or more players. Na et al. use the noncooperative Nash and Stackelberg game theory to model the human-automation interaction for obstacle avoidance [83, 84]. Flad et al. propose the differential game framework between the human driver and Lane Keeping Assistance (LKA) system to decrease conflicts [85]. The work in [86] adopts the stochastic game theory to address human-automation conflicts by considering driver model's uncertainties.

1.4   SUMMARY

Considering that the optimal control method is a non-convex optimization problem for the obstacle avoidance problem in a complex environment. When the system has more parameters to solve, it will fall into a dimensional curse and the computational consumption will increase significantly, which is unacceptable for the intelligent vehicle once the computation resource is limited. Therefore, we generally decomposed this trajectory optimization problem into the path (speed)-planning and path (speed) following. For instance, when in the phase of the path planning, it aims to generate a safe, collision-free, and dynamic feasible route to reach the goal position during the traveling. This is based on the information of road, static or dynamic obstacles around the vehicle as inputs, meanwhile, the path subjects to multiple dynamical constraints.

As introduced above, we may notice that most path-planning algorithms are usually originated from the applications of mobile robotics; they need proper modifications before applying to the field of intelligent vehicles. For example, the conventional potential method is difficult to deal with the path oscillations in the narrow channel, as well as the local minima trap, also, it cannot guarantee the planned path with satisfying curvatures. Similarly, the A* algorithm is usually based on the environment map consisting of grids which brings inconveniences to the smoothness of the path curvature. The sampling-based RRT algorithm generally lacks repeatability due to random search, and its parameter selection often needs to be adjusted for different scenarios. Intelligent swarm algorithms such as the ACO, IWD, etc., might need a long optimal solution-finding time with a slow convergence speed, which fails to meet the high requirements of the path planning for intelligent vehicles in real time.

More importantly, there exist remarkable differences between intelligent vehicles and mobile robots in path planning. On the one hand, the speed of the robotic is usually low, thus, there is no need to consider the vehicle stability in the path-planning procedure while avoiding obstacles; however, it needs extra attention for the intelligent vehicle path-planning at high speeds, which demands the curvature of the planned path be appropriate to preventing potential dangers (e.g., instability). In addition, the velocity also needs to be carefully designed in the stage of speed-planning. On the other hand, there are usually no road boundary constraints in the configuration space of mobile robotics, still it is necessary to keep the ego vehicle always within the roads, which generally leads to an increase in the cost of solution finding. Therefore, there are still improvements for trajectory-planning (path-planning, speed-planning) algorithms applied for intelligent vehicles when motivated by the robotic path-planning applications. Consider that the potential method for the path planning has been studied well, which has a simple structure and is convenient for bottom-level control, and the combination of the elastic band improves the quality of the path curvature. We will further enhance the potential of the elastic band with an improved hazard potential construction. Furthermore, we also target an optimal path-planning method in real time for the intelligent vehicle, which is based on the technique of the convex optimization and natural cubic splines. This will be expanded upon in subsequent chapters.

As for the path following, although most methods can cope with simple following tasks based on the linear 2DOF vehicle model. Nevertheless, the model parameters of the vehicle system may be inaccurate in practice and it is necessary to enhance the robustness of the tracking algorithm. Moreover, the 2DOF model has a disadvantage which is that a strong assumption needs to be held where the moving speed of the vehicle shall remain constant, and the linear lateral tire model is adopted. This simple model can no longer meet the requirements when a dynamic longitudinal motion is involved or the vehicle tires run in a nonlinear working condition (e.g., high lateral accelerations). It is necessary to develop a trajectory-following controller based on a more complex vehicle model. Therefore, on one hand, we will introduce a robust adaptive Sliding Mode Control (SMC) method based on nonlinear vehicle models for the velocity-tracking and the path-tracking. On the other hand, we will introduce an alternative approach to realize the trajectory-tracking (profiles of the velocity and the path) with a linearized time-varying MPC controller based on a nonlinear 5DOF vehicle model and convex optimization technique.

Regarding for the human-automation cooperative driving strategy, prior work has done great research on the control authority and conflict management, however, there exist some aspects to be improved. Instead of driver's actions or states, the driver's driving intention recognized by machine learning methods is directly considered into the human-automation cooperative driving strategy to understand driver's desired maneuver and decrease control authority conflicts. Thus, we will introduce the data collecting experiment and machine learning approaches for driving intention recognition. Support vector machine (SVM), that is a supervised machine learning method, and an inductive multi-label classification with an unlabeled data (iMLCU), that is a semi-supervised method will be used to predict driving intention. It is also important to consider the driving risk when the design of the human-automation cooperative driving strategy. Driving safety field and TTC are employed into the situation assessment of the strategy. In our study, there are two control strategies for human-automation cooperative driving systems proposed based on the driving intention or driving risk. In the first strategy, the non-cooperative Nash game theory is used to model the interaction between path-following controller and human drivers, and a dynamic authority allocation strategy is designed based on the driving risk evaluated by driving safety field to adjust control authorities between two participants. Another strategy is proposed by using fuzzy control theory. In this strategy, the driver's driving intention, evaluated driving risk, and performance index are taken into consideration for the fuzzy-based cooperative driving strategy.

CHAPTER 2

Decision Making for Intelligent Vehicles

2.1 INTRODUCTION

Decision making for intelligent vehicles is popular in recent years, which plays an important part in the intelligent vehicle automatic driving system. It also acts as a bridge connecting the upper layer of the environment perception and the lower level of the planning and control.

On the one hand, from the perspective of the overall architecture, as depicted in Figure 2.1, existing decision-making methods for intelligent vehicles mainly include two categories, namely, the hierarchical decision making, and the end-to-end decision making. The hierarchical decision making is applied in the typical traditional architecture of the perception-decision-planning and control. More specifically, the hierarchical decision making determines proper tactical maneuvers based on the information of the vehicle and its surrounding environments, and then send it to the planning and control layers for specific trajectories and corresponding control orders, finally, the commanding orders will be handed over to the actuators (steering/throttle/control, etc.) of the vehicle systems for executions. The end-to-end decision-making which is based on the raw sensor information decides the control inputs to the vehicle directly and sends them to actuators for executions in real time.

On the other hand, from the perspective of the decision-making mechanism, existing decision-making methods for intelligent vehicles could mainly be divided into two categories, namely, the decision making driven by artificial setting mechanisms and the one driven by the data. The former category is based on certain artificially predefined decision-making mechanisms (such as rule-based, target-function based, etc.) to realize the decision making, typical methods include the Finite State Machine (FSM), Dynamic Game, etc. While the data-driven decision-making methods, instead, learn decision-making strategies spontaneously from the data, typical methods include imitation learning and reinforcement learning, etc.

2.2 DECISION-MAKING METHODS

This section will introduce several popular existing decision-making methods, including the FSM, Dynamic Game, imitation learning, and the enforcement learning.

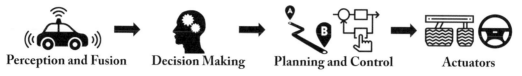

Perception and Fusion   Decision Making   Planning and Control   Actuators

(a) The hierarchical decision making

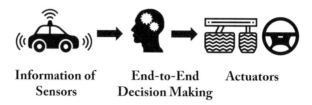

Information of   End-to-End   Actuators
Sensors   Decision Making

(b) The end-to-end decision making

Figure 2.1: The flowchart of the decision making.

2.2.1   FINITE STATE MACHINE FOR DECISION MAKING

The rule-based FSM is one of the earliest applications of decision-making methods for intelligent vehicles. Generally speaking, there are several finite discrete states of the vehicle which are artificially defined in the FSM (as shown in Figure 2.2); the transition conditions and transition actions among states are specified to form the rule database (e.g., using if-then; else-if-then). Therefore, the FSM would first confirm the current state based on the information of environmental perception when in a decision-making process, then query the state transition conditions in the rule database to determine the next target state and the corresponding state transition action. Finally, the state transition action regarded as the output of the system.

Overall, the FSM is simple in principle, with low computation source requirements, and also easy to be implemented on the in-vehicle platform. However, the FSM has limitations and the complexity of the FSM rules increases exponentially when the state space is large, which affects the efficiency of the rule querying. Although the hierarchical FSM which makes decisions from top to bottom are often applied in practice, it is convenient to setting rules, as well as improving query efficiency. Still, it is difficult to handle high-dimensional problems, which leads to low decision accuracy in complex scenarios. Besides, the FSM is difficult to adapt to the complicated and changing road environment, because it requires to manually set rules for state transitions which is difficult to be completed for all situations, or worse, failures of decision making might occur when it encounters an out-of-rule situation while seriously affecting the driving safety. Therefore, the application of the FSM is only limited to simple scenarios (such as closed industrial parks).

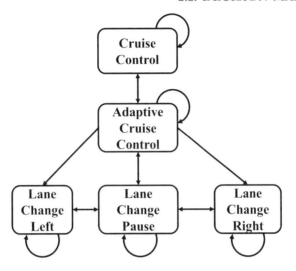

Figure 2.2: The hierarchical FSM.

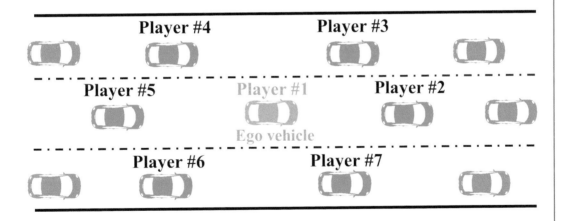

Figure 2.3: The illustration of decision making by dynamic game.

2.2.2   DYNAMIC GAME FOR DECISION MAKING

As depicted in Figure 2.3, the dynamic game method for decision-making treats the ego-vehicle and its surrounding vehicles as game participants through a cooperative or non-cooperative dynamic game approach. First, we had to define the strategy space of each game participant and its corresponding payoff function, and then solve the equilibrium points of the game while satisfying corresponding constraints to obtain decision results.

The decision making for intelligent vehicles by the dynamic game draws on the internal decision-making mechanism of the human driver while driving, and takes into account the dynamic influence of the surrounding vehicles while making decisions. However, the Nash game [87] and Stackelberg game [88], which are commonly used for the game, are both simplifications of the real situation, and there may be inevitable deviations in the prediction of the surrounding vehicles' behaviors. Moreover, the manually predefined payoff function of each game participant will highly affect the decision results, and one has to find the optimal payoff function with trials. Further, the calculation burden of solving the game equilibrium point is huge when there are too many game participants, or the payoff functions and constraints of the game participant are complex. Therefore, decision making by dynamic game methods might be difficult to be implemented online in real time.

2.2.3   IMITATION LEARNING FOR DECISION MAKING

The basic premise of the imitation learning for decision making is to assume that the human driver can be fully competent for the decision making; therefore, the decision maker via imitation learning can also be competent for the task if it follows the human driver strategy. Based on the data collection of human drivers in various driving scenarios through driving experiments or driving simulator experiments, and the observation-behavior pairs: $D = \{\langle o_i, a_i \rangle\}_{i=1}^{N}$ are extracted, where o_i denotes the sample observations and a_i denotes the behavior of the driver in the driving scenario. Then a supervised learning method is used to construct the decision maker, and model training is performed based on the sample data, while on the premise of ensuring the generalization ability, the difference between the output of the decision-maker, and the output of the human driver is minimized, namely,

$$\min_{\theta} \sum_{i} J(F(o_i; \theta), a_i), \tag{2.1}$$

where $F(o_i; \theta)$ denotes the output of the decision-maker, a_i denotes the output of the human driver, and J denotes the error evaluation function.

A large number of scholars have conducted in-depth research on the supervised learning, and many mature and reliable methods are available, such as classification regression models, SVM, decision trees, deep neural networks (DNN), etc. The imitation learning for decision making has sufficient theoretical basis and algorithm reserves, with high feasibility for implementation. However, while imitating the expert driver strategy, the ability of the decision maker cannot exceed the human driver. Since the method is data-driven, multiple rounds of iterative testing are required to fully verify the safety and robustness of the algorithm; as a consequence, the implementation cost is high.

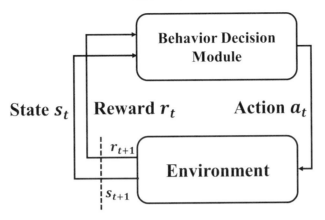

Figure 2.4: The Markov decision process.

2.3   DECISION MAKING BY DEEP Q-LEARNING

As introduced above, decisio making by reinforcement learning methods has promising prospects for the application in intelligent vehicles, although they are still immature and require further theoretical researches. In this section, we will propose a DQN-based decision-making method.

2.3.1   MARKOV DECISION PROCESS

The decision making of the intelligent vehicle is quite complicated as there might be many random factors that are related to the surrounding environment, and the problem needs to be simplified to obtain feasible solutions. One assumption is made that the decision is only related to the current observation states, while no relationships with the historical observation states, such that, it can be constructed as a Markov decision process $\langle S, A, R, \gamma \rangle$, where S denotes the observation state space and A denotes the decision in the action space, while R denotes the reward and γ denotes the discount factor. Considering high randomness in a dynamically changing environment, the state transition probability is supposed to be unknown; the Markov process could be solved by using a model-free approach.

As shown in Figure 2.4, the refinement decision-making module interacts with the environment online under the guidance of the reward value. More specifically, it selects the decision action as output $a_t \in A$ based on the observation states of the ego-vehicle and its surrounding vehicles' movement at the time step t. The reward function evaluated at the current time $r_t = S \times A \rightarrow R$ and the updated observation state S_{t+1} with the environment feedbacks are obtained once the execution of the action is completed. Therefore, the solution of the Markov decision-making process is identical to actively exploring possible strategies from $\pi : S \rightarrow A$ to find an optimal strategy π^* that maximizes the expectation of the discount reward within the period T, namely,

$$\pi^* = \max_\pi \mathbb{E}_{\tau(\pi)} \left[\sum_{t=0}^T \gamma^t r_t \right], \tag{2.2}$$

where $\tau(\pi)$ denotes the decision trajectory under the strategy π, r_t denotes the reward value with the environmental feedback at time t, $\mathbb{E}$ denotes the expectation operator, and $\gamma \in [0, 1]$ denotes the discount coefficient which reflects the importance of short-term rewards compared to long-term rewards, meaning it values long-term rewards more if γ approaches more closely to 1.

2.3.2   DEEP Q-NETWORK (DQN)

The DQN is a value-based, off-policy reinforcement learning method, which is the combination of reinforcement learning and DNNs. It can solve a model-free Markov decision-making process with limited and discrete action space. The basic idea of DQN is to estimate the value of each decision action in the observation state through the DNNs called the Q-network, and mechanisms such as the ε-greedy and the Noisy Net are applied to choose the best action based on the output of Q-network. The key mechanism of DQN is the experience replay, which stores the decision-making experience every time it interacts with the environment, such that we could train the Q-network and optimize the estimation of the action value by revisiting and batch sampling history experience in the experience replay library. A brief flow work of the DQN is shown in Table 2.1.

Remark 2.1   To balance the exploration and utilization of strategies, the initial exploration rate is often set to a large value (such as 0.9) to ensure sufficient strategy explorations are carried out in the early stage of learning when applying the DQN in practice. As the learning process continues, the exploration rate is then gradually reduced which aims to fully take advantage of the optimal strategies that have been found, more importantly, to ensure the convergence of strategy learning.

Remark 2.2   There are a variety of improved DQN variants, such as the Double DQN [89] that can effectively alleviate the overestimation problem, and the Prioritized Replay DQN [90] which improves the efficiency of the experience playback sample, the Dueling DQN [91] which splits the state-action value into the state value and the advantage of actions to speed up the learning process, and the Rainbow DQN [92] which integrates a variety of improved mechanisms such as. However, like the original DQN, the above methods can be only used for decision making that the action space is finite and discrete, and cannot deal with the problem of high-dimensional or continuous state space.

Algorithm 2.1 Pseudo codes of the DQN

Input: Observation states of the ego vehicle and surrounding vehicles.

Initialization: The structure and training parameters of the Q-network, the maximum iteration T, the exploring rate ε, the experience replay sample quantity k, the capacity of experience replay library C, the corresponding number of samples in experience replay m_0 which can start the learning phase.

Output: The decision action a_i.

1: Randomly initialize the parameter of the Q-network w, such that the Q-network can randomly estimate the state-action value.

2: Clear the experience Replay library D.

3: **for** $i = 1$ to T **do**

4:      Enter the observed state s_i into the Q-network to obtain the state-action value for all actions, then select the best action ai via the ε-greedy algorithm.

5:      Perform the action a_i to get the updated observation state s'_i, the feedback reward value r_i and the termination status flag is_end_i (whether the termination state is reached).

6:      Store the interactive experience $\{s_i, s'_i, a_i, r_i, \text{is}_e\text{nd}_i\}$ into the experience Replay library.

7:      Update observation status by $s_{i+1} = s'_i$.

8:      **if** $i > m_0$ **then**

9:          Train the Q-network, calculate the target Q-value y_j by randomly sampling m samples from the Experiment Replay library $\{s[j], a[j], r[j], s'[j], \text{is}_e\text{nd}[j]\}$, $j = 1, 2, \ldots, m$

$$y_j = \begin{cases} r_j, & \text{is}_e\text{nd}[j]\,\text{isTRUE} \\ r_j + \gamma \max_a Q(s_j, a_j, w), & \text{is}_e\text{nd}[j]\,\text{isFALSE} \end{cases}$$

10:     **end if**

11:     Use the MSE loss function $LOSS = \frac{1}{m} \sum_{j=1}^{m}(y_j - !(s_j, a_j, w))^2$ to calculate the loss value and update parameters of the Q-network by the gradient backpropagation algorithm.

12:     **if** s'_i is the terminate state **then**

13:         Break.

14:     **else**

15:         Continue.

16:     **end if**

17: **end for**

Figure 2.5: The grayscale image obtained from the forwarding vision sensor.

2.3.3 CASE STUDY

This section implements an end-to-end decision-making demo based on the aforementioned DQN reinforcement learning. This demo uses the open-source Unity ML-Agents virtual driving environment by Kyushik Min, which is a suburban highway driving scenario. Please refer to [93, 94] and the Open source project on Github by Kyushik Min for a detailed introduction to the virtual driving environment. The ego vehicle is equipped with forwarding vision sensors (the output grayscale image is shown in Figure 2.5) and the lidars on the top, the output frequency of the sensors can be customized, as well as the number of environment vehicles in the scenario, the visibility status, etc.

The observation states of the decision are the original gray image obtained from the forwarding vision sensor and the raw point cloud data from Lidar, and the decision action space is set to five discrete actions, namely lane-keeping with constant speed, lane-keeping with accelerating, lane-keeping with decelerating, left lane-change with constant speed, and right lane-change with constant speed. The Q-network of the DQN consists of three layers of convolutional layers where ReLU is chosen as the activation function and one fully connected layer with ReLU as the activation function. The decision action is selected through the ε-greedy algorithm according to the predicted state-action value evaluated by the Q-network. The parameter settings of policy learning are shown in Table 2.1.

The reward function defined in Equation (2.3) includes the reward from the longitudinal and lateral behaviors which considers the traffic efficiency and safety simultaneously.

$$Reward = Longitudinal_{reward} + Lateral_{reward} + Overtake_{reward} + Collision_{reward}, \qquad (2.3)$$

where

Table 2.1: The parameter settings of reinforcement learning

Parameter	Value
Total training steps	1e6
The learning rate	2.5e-4
Mini-batch size	32
The discount factor γ	0.99
The initial value of ε	1
The final value of ε	0.1
The change rate of ε	9e-7
Replay memory size	1e5

- the $Longitudinal_{reward}$ is the difference between the expected speed and the current speed,

$$Longitudinal_{reward} = \frac{speed - speed_{\min}}{speed_{target} - speed_{\min}}; \tag{2.4}$$

- the $Lateral_{reward}$ is the penalty for lane change, such that avoiding unnecessary frequent lane-changing to keep satisfying driving comforts.

$$Lateral_{reward} = \begin{cases} -0.4, & \text{if lane} - \text{change} \\ 0, & \text{otherwise}; \end{cases} \tag{2.5}$$

- the $Overtake_{reward}$ is proportionate to the change of the number of overtaken vehicles after the decision is made, which encourages the ego vehicle to overtake actively to improve traffic efficiency

$$Overtake_{reward} = 0.5 \times (N_{overtake} - N_{0,overtake}), \tag{2.6}$$

where $N_{overtake}$ denotes the number of overtaken vehicles after the action is executed, $N_{0,overtake}$ denotes the number of overtaken vehicles before the action is executed; and

- the $Collision_{reward}$ which targets driving safety is the penalty for the collisions between the ego vehicle and surrounding vehicles and lane boundaries.

$$Collision_{reward} = \begin{cases} -4, & \text{if Collision happens} \\ 0, & \text{otherwise}. \end{cases} \tag{2.7}$$

The DQN decision-making module is constructed based on the Tensorflow deep learning framework, and the above-mentioned virtual driving environment is used to carry out the

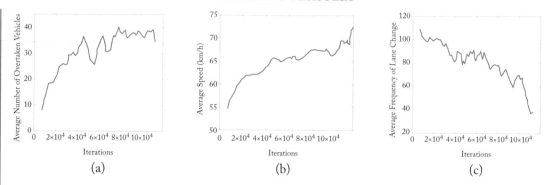

Figure 2.6: Model training results of the ego vehicle per five training episodes: (a) average speed per five-episodes. (b) average frequency of lane-change per five-episodes; and (c) average numbers of overtaken vehicles per five-episodes.

Table 2.2: Scenario verification comparison results between the DQN and the FSM

Parameter	DQN	FSM
Cumulative number of collisions	0	0
Average velocity (km/h)	72.1	63.5
Standard deviation of speed (km/h)	5.6	10.3
Average lane-change frequency in a single-task	8	4

reinforcement learning. During the training process, the average speed, average lane change times, and average numbers of overtaking vehicles by the ego vehicle per five training episodes are shown in Figure 2.6. It can be seen that the DQN decision-making module optimizes its decision-making policy gradually through policy exploration and utilizations in the interaction with the virtual driving environment, the policy's traffic efficiency and smoothness are improved steadily, and the policy learning tends to converge after 100,000 training episodes.

To verify the performance of the DQN decision-making method, it is compared by the FSM method designed by reference [95]. The above two decision-making methods are used to control the host vehicle to complete five point-to-point automatic driving tasks in the above-mentioned virtual driving environment, with the driving distance being about 4 km at each time. During the test, the ego vehicle's cumulative collision times, average velocity, the standard deviation of velocity, average lane change times of single-task are counted, and the results are summarized in Table 2.2. It can be observed that the decision-making policy of the DQN decision-making method seems more active than that of the FSM decision-making method: the DQN improves the traffic efficiency by actively adopting safe lane change behavior, and reduces the standard deviation of vehicle speed by avoiding conservative emergency braking, which achieve better balances among safety, efficiency, and driving comforts.

2.4 SUMMARY

We simply present a novel DQN decision-making method that can learn an ideal end-to-end decision-making policy through interactive training in this chapter, and the simulation results show that the performance seems better than that of the traditional FSM method. The deep reinforcement learning method shows promising potentials for intelligent vehicle decision-making tasks. However, the situation of incomplete and inaccurate observation state caused by sensor noise is not considered in the current work to simplify the problem, while that will be further considered to improve the robustness of the algorithm in our future work.

Path and Speed Planning for Intelligent Vehicles

In this chapter, we will introduce our latest research work on the path-planning and speed-planning algorithms for intelligent vehicles. The path-planning methods covered in this book mainly include the elastic band and its modifications based on the well-known artificial potentials, besides the optimal path planning based on the combination of the convex optimization and the natural cubic splines are introduced to obtain optimal path profiles for different driving maneuvers (e.g., the free driving, lane-keeping, obstacle avoidance). Similarly, the speed planning for non-following cases can be realized by the convex optimization or multi-objects optimization based on high-order polynomials or natural cubic splines, while the MPC method will be developed for the speed planning in car-following scenarios.

3.1   PREMIERE

Before heading to the implementation of path-planning algorithms, we will introduce the geometric descriptions on the structured road first, followed by the definitions and useful properties on the harmonic function and natural cubic spline, which are used for the path planning of intelligent vehicles.

3.1.1   GEOMETRIC MODEL OF THE STRUCTURED ROAD

Information on the road environment can be obtained by Lidar or vision-based active safety products such as MobiEye®. Suppose we choose the origin of the inertial Cartesian coordinate system at the position on the road centerline that corresponds to the longitudinal center of gravity of the vehicle at the very beginning of the path planning. And assuming the global coordinate position of the road centerline $O' = \{X_c, Y_c\}$ is a set of parametric function f_X, f_Y respecting to the arc length (distance) of the road centerline s_c, namely

$$\begin{cases} X_c = f_X(s_c) \\ Y_c = f_Y(s_c). \end{cases} \tag{3.1}$$

Besides, if the road centerline is made up of a series of way-points (size of N), the distance along the centerline which is evaluated at each way-point can be approximated by

$$s_{c,i} = s_{c,i-1} + \sqrt{(X_{c,i} - X_{c,i-1})^2 + (Y_{c,i} - Y_{c,i-1})^2}, \tag{3.2}$$

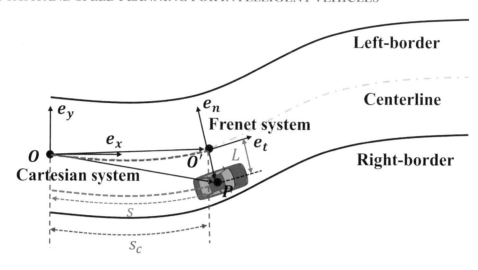

Figure 3.1: Illustration of the coordinate system.

where $i = 1, \ldots, N, s_{c,0} = 0$.

As shown in Figure 3.1, the tangent unit vector e_t and the nominal unit vector e_n, also called the unit vectors of the Frenet coordinate system, can be calculated by

$$
\begin{cases}
e_t(s_c) = \dfrac{f_X'(s_c)}{\sqrt{f_X'^2(s_c) + f_Y'^2(s_c)}} e_x + \dfrac{f_Y'(s_c)}{\sqrt{f_X'^2(s_c) + f_Y'^2(s_c)}} e_y \\[3mm]
e_n(s_c) = \dfrac{-f_Y'(s_c)}{\sqrt{f_X'^2(s_c) + f_Y'^2(s_c)}} e_x + \dfrac{f_X'(s_c)}{\sqrt{f_X'^2(s_c) + f_Y'^2(s_c)}} e_y,
\end{cases}
\tag{3.3}
$$

where e_x and e_y are the unit vector of the earth coordinate, respectively. If one arbitrary position $P = (X, Y)$ on the road, the corresponding centerline distance s_c and the lateral distance L to the road centerline are already known, the pair (s_c, L) is called the coordinate of the Frenet system of the road. According to the relationship shown in Figure 3.1, $\vec{OP} = \vec{OO'} + \vec{O'P} = \vec{OO'} + Le_n$, such that

$$
\begin{cases}
X = f_X(s_c) - \dfrac{L f_Y'(s_c)}{\sqrt{f_X'^2(s_c) + f_Y'^2(s_c)}} \\[3mm]
Y = f_Y(s_c) + \dfrac{L f_X'(s_c)}{\sqrt{f_X'^2(s_c) + f_Y'^2(s_c)}}.
\end{cases}
\tag{3.4}
$$

For simplicity, it yields

$$
R = Z(\mathcal{F}) = (z_1(R), z_2(R)),
\tag{3.5}
$$

where $Z = (z_1, z_2)$ represents the mapping function between the Frenet coordinate and the world coordinate as shown in Equation (3.4), such that the inverse function of Z can be expressed by

$$\mathcal{F} = z^{-1}(R) = (z_1^{-1}(R), z_2^{-1}(R)). \tag{3.6}$$

3.1.2   HARMONIC ARTIFICIAL POTENTIAL FUNCTION

The harmonic function is a second-order continuously derivable function $f : U \rightarrow R$, where U is an open subset of R^n, and f satisfies the Laplace equation, namely

$$\sum_{i=1}^{n} \frac{\partial^2 f_i}{\partial u_i^2} = 0, \tag{3.7}$$

where Equation (3.7) is generally written as

$$\triangle f = 0, \tag{3.8}$$

where $\triangle$ represents the operator of the Laplace.

We will find that the harmonic function has several friendly and useful properties, which is of great convenience to the application of the artificial potential in path planning. Useful properties will be listed as follows.

(1) The result of the sum, difference, and multiplication of the harmonic function is still harmonic function, which means that if the dangerous potential fields represented by the harmonic function are superimposed. The result is still a harmonic function.

(2) If f is a harmonic function on U, then all partial derivatives of f are still harmonic functions on U.

(3) If K is a compact subset of U, then the function induced by f on K can only reach its maximum and minimum values on the boundary. Especially, if U is connected, then the non-constant function f cannot reach the maximum and minimum values. This property is often called the Maximum Value Theorem of the harmonic function. If we choose the harmonic function as a candidate for the repulsive function of the obstacle potential field, obviously, its maximum value appears at the boundary of the obstacle, thereby avoiding the local minimum trap in conventional artificial potential methods.

3.1.3   NATURAL CUBIC SPLINE

Generally speaking, a cubic spline curve is a piecewise cubic polynomial curve composed of a series of control points (e.g., $n + 1$ control points in total). If the second derivative at the end node of the curve is zero, it is the so-called natural cubic spline. Since spline coefficients can be obtained by solving linear equation sets, thus it is widely used. We will use natural cubic splines to describe the planned path in this book. At first, let us consider a set of one-dimensional spline curves consisting of $n + 1$ points and one spline curve Y_i, $1 \leq i \leq n$ is described by a cubic polynomial, as shown in Equation (3.9),

$$Y_i(t) = a_i + b_i t + c_i t^2 + d_i t^3, t \in [0, 1], \qquad (3.9)$$

where a_i, b_i, c_i, d_i denote unknown coefficients that needed to be further derived and t is a parametric variable of the spline varying from 0 to 1.

According to the known values of the beginning and end node for each segment, we can get

$$\begin{cases} Y_i(0) = y_i = a_i \\ Y_i(1) = y_{i+1} = a_i + b_i + c_i + d_i, \end{cases} \qquad (3.10)$$

where y_i denotes the value evaluated at the ith control point.

Differentiating Y_i to t and evaluated at the control point further gives

$$\begin{cases} Y_i'(0) = b_i = D_i \\ Y_i'(1) = b_i + 2c_i + 3d_i = D_{i+1}, \end{cases} \qquad (3.11)$$

where D_i is a newly introduced solution variable. Combining Equations (3.10)–(3.11), a_i, b_i, c_i, d_i are obtained:

$$\begin{cases} a_i = y_i \\ b_i = D_i \\ c_i = 3(y_{i+1-y_i} - (2D_i + D_{i+1})) \\ d_i = 2(y_i - y_{i+1}) + D_i + D_{i+1}. \end{cases} \qquad (3.12)$$

Furthermore, considering that the second derivative at the end of each segment of the spline should be consistent, thus,

$$\begin{cases} Y_{i-1}(1) = y_i \\ Y_{i-1}'(1) = Y_i'(0) \\ Y_{i-1}''(1) = Y_i''(0) \\ Y_i(0) = y_i. \end{cases} \qquad (3.13)$$

For the first and last points of the curve,

$$\begin{cases} Y_0(0) = y_0 \\ Y_{n-1}(1) = y_n. \end{cases} \qquad (3.14)$$

Equations (3.13) and (3.14) can only obtain $4(n - 1) + 2 = 4n - 2$ equations while with a total of $4n$ unknowns, thus, two additional conditions need to be obtained. Indeed, we may notice that the second derivative at the first and last endpoints of the curve is zero, namely,

$$\begin{cases} Y_0''(0) = 0 \\ Y_{n-1}''(1) = 0. \end{cases} \qquad (3.15)$$

In this way, combining Equations (3.13)–(3.15), we can get a system of linear equations, as shown in Equation (3.16):

$$
\begin{bmatrix}
2 & 1 & & & & & \\
1 & 4 & 1 & & & & \\
& 1 & 4 & 1 & & & \\
& & 1 & 4 & 1 & & \\
\vdots & \ddots & \ddots & \ddots & \ddots & \ddots & \ddots \\
& & & & 1 & 4 & 1 \\
& & & & & 4 & 1
\end{bmatrix}
\begin{bmatrix}
D_0 \\ D_1 \\ D_2 \\ D_3 \\ \vdots \\ D_{n-1} \\ D_n
\end{bmatrix}
= 3
\begin{bmatrix}
(y_1 - y_0) \\ (y_2 - y_1) \\ (y_3 - y_1) \\ \vdots \\ (y_{n-1} - y_{n-3}) \\ (y_n - y_{n-2}) \\ (y_n - y_{n-1})
\end{bmatrix}.
\tag{3.16}
$$

Set $D = [D_0, D_1, \ldots, D_n]^T$,, the above linear equation set is expressed in a matrix form, such that

$$
\mathbf{M}D = 3\mathbf{N}y,
\tag{3.17}
$$

where $\mathbf{M}, \mathbf{N}$ are constant matrices and $\mathbf{N}y = \mathbf{matrix}[y_i - y_{i-1}], i = 1, \ldots, n$. Once D is confirmed, according to (3.11) and (3.11), it yields

$$
\left. \frac{dY(t)}{dt} \right|_{t=0} = D = (3\mathbf{M}^{-1}\mathbf{N})y
\tag{3.18}
$$

$$
\left. \frac{d^2Y(t)}{dt^2} \right|_{t=0} = 6\mathbf{N}y - 2P - D = 6(\mathbf{N} - P\mathbf{M}^{-1}\mathbf{N})y,
\tag{3.19}
$$

where $PD = \mathbf{matrix}[2D_i + D_{i+1}], i = 0, \ldots, n-1$ and P is a constant matrix as well. Rewriting Equations (3.18)–(3.19) gives

$$
\begin{cases}
\left. \dfrac{dY(t)}{dt} \right|_{t=0} = \mathbf{G}y \\[2ex]
\left. \dfrac{d^2Y(t)}{dt^2} \right|_{t=0} = \mathbf{H}y,
\end{cases}
\tag{3.20}
$$

where $\mathbf{G} = 3\mathbf{M}^{-1}\mathbf{N}, \mathbf{H} = 6(\mathbf{N} - P\mathbf{M}^{-1}\mathbf{N})$, which means the first and second derivatives of the natural cubic spline at each node can be expressed as a linear form respecting the control points y. This property is extremely useful to construct quadratic terms related to path curvatures.

3.2 PATH PLANNING WITH ELASTIC BANDS

For the vehicle local path planning, the design of an ideal artificial potential should meet the following criteria:

(1) The construction of the hazardous potential regarding road boundaries must keep the host vehicle always within the road. Therefore, we usually use a harmonic function (such as the logarithmic function or a negative exponential function) to build the road boundary potential, and because of that its potential value at the boundary tends to be infinity.

(2) The directional hazard description of the obstacle is expected to be related to the longitudinal gap between the host vehicle and the obstacle other than in lateral direction. That means it is far safer for the host vehicle to stay 1 m away from the obstacle in its lateral direction than to keep 1 m away from the obstacle in its longitudinal direction.

Therefore, we shall carefully pay attention to the potential setting of the obstacle in its longitudinal direction. This chapter will propose a novel hazardous potential called the **guiding obstacle potential** through a scale of the longitudinal gap to enlarge its influence, such that the vehicle maintains a sufficiently safe distance to the obstacle when performing obstacle avoidance maneuvers. For simplicity, let us go through this harmonic potential path planning with elastic bands to deal with the static obstacle first.

3.2.1 HAZARD POTENTIAL CONSTRUCTION

As mentioned above, we select the harmonic function as a candidate for the hazardous potential regarding road boundaries or obstacles, which aims to ensure the host vehicle stay away from obstacles, meanwhile, keep the host vehicle moving within road boundaries. The construction of these hazardous potentials including road boundary potentials, obstacle potentials, as well as the novel guiding potential, will be introduced below.

Based on the experience of the artificial potential field model in the field of robotics, a virtual potential represented by Equation (3.21) was selected for the road boundary potential. When the vehicle approaches the road boundary, its potential energy tends to be infinity, so as to prevent the vehicle from approaching the boundary, which is also depicted in Figure 3.2.

$$U_{border} = -k_{border,l} \ln |b - L| - k_{border,r} \ln |b + L|, \tag{3.21}$$

where $k_{border,l}$ and $k_{border,r}$ denote the stiffness of the road left and right boundary, respectively, b denotes the road lane width, and L denotes the Frenet coordinate vector corresponding to $\boldsymbol{R}$.

Based on the relationship of Equation (3.4) with the world coordinate system and the Frenet coordinate system, $\boldsymbol{R} = \boldsymbol{Z}(\mathcal{F}) \Rightarrow \mathcal{F} = \boldsymbol{Z}^{-1}(\boldsymbol{R})$, where $\boldsymbol{Z}(\cdot)$ represents the mapping function between the Frenet coordinate system and the world coordinate system, while $\boldsymbol{Z}(\cdot)$ denotes the inverse mapping function of $\boldsymbol{Z}$.

According to Equation (3.4),

$$L = \sqrt{1 + \left(\frac{f_Y'(\boldsymbol{Z}^{-1}(\boldsymbol{R})e_t)^2}{f_X'(\boldsymbol{Z}^{-1}(\boldsymbol{R})e_t)^2} \right)^2} \left(y - f_Y'(\boldsymbol{Z}^{-1}(\boldsymbol{R})e_t) \right), \tag{3.22}$$

where $e_x = (1, 0)$, $e_y = (0, 1)$ denotes the unit vector in the world coordinate system. Then $\boldsymbol{Z}^{-1}(\boldsymbol{R})e_x$ equals the longitudinal distance component of the Frenet coordinate $\mathcal{F}$ while given the world coordinate $\boldsymbol{R}$. Besides, the world coordinates of the road centerline (x_c, y_c) are evaluated by functions respecting the longitudinal distance along with road centerline s_c, namely,

$$f_X : s_c \rightarrow x_c, f_Y : s_c \rightarrow y_c.$$

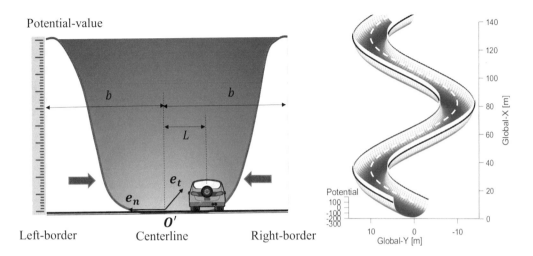

Figure 3.2: Illustration of the road borders.

Substituting (3.22) into (3.21) yields the completely form of the road border potential

$$U_{border} = -k_{border,l} \ln \left| b - \sqrt{\left(\frac{f_Y'(\mathbf{Z}^{-1}(\mathbf{R})\mathbf{e}_x)^2}{f_X'(\mathbf{Z}^{-1}(\mathbf{R})\mathbf{e}_x)^2} \right)^2 \left(\mathbf{R}\mathbf{e}_y - f_Y'(\mathbf{Z}^{-1}(\mathbf{R})\mathbf{e}_x) \right)} \right|$$
$$- k_{border,r} \ln \left| b + \sqrt{\left(\frac{f_Y'(\mathbf{Z}^{-1}(\mathbf{R})\mathbf{e}_x)^2}{f_X'(\mathbf{Z}^{-1}(\mathbf{R})\mathbf{e}_x)^2} \right)^2 \left(\mathbf{R}\mathbf{e}_y - f_Y'(\mathbf{Z}^{-1}(\mathbf{R})\mathbf{e}_x) \right)} \right| . \tag{3.23}$$

Figure 3.2 also shows an example of the road border potentials of a curved road.

Similarly, as shown in Figure 3.3a, the establishment of the obstacle potential is to keep the host vehicle beyond a certain safety distance away from the obstacle. Thus, the corresponding potential energy value U_{obst} in Equation (3.24) needs to be dramatically large when near the boundary of the obstacle,

$$U_{obst} = \begin{cases} -k_{obst} \sum_{m=1}^{M} (\ln |\|\mathbf{R} - \mathbf{R}_{o,m}\| - r_{sc}|, & \text{if } \|\mathbf{R} - \mathbf{R}_{o,m}\| > r_{sc}) \\ +\infty, & \text{otherwise,} \end{cases} \tag{3.24}$$

where k_{obst} denotes the stiffness of the guiding potential, M denotes the total number of obstacles in the detecting range, m denotes the sequence index of the obstacle, r_{sc} denotes the radius of the safety circle, and $\mathbf{R}_{o,m} = (x_{o,m}, y_{o,m})$ denotes the world coordinate of the mth obstacle's center position, given the Frenet coordinate of the mth obstacle $(s_{co,m}, L_{o,m})$. Again, based on

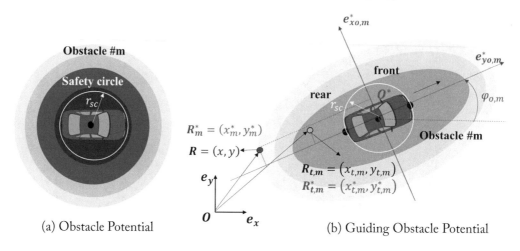

(a) Obstacle Potential   (b) Guiding Obstacle Potential

Figure 3.3: Illustration of the obstacle potential and the guiding potential.

Equation (3.4), $\boldsymbol{R}_{o,m} = (x_{o,m}, y_{o,m})$ can be calculated via

$$\boldsymbol{R}_{o,m} = \left[f_X(s_{co,m}) - \frac{L_{o,m} f'_Y(s_{co,m})}{\sqrt{f'^{(2)}_X(s_{co,m}) + f'^2_Y(s_{co,m})}} + \frac{L_{o,m} f'_X(s_{co,m})}{\sqrt{f'^{(2)}_X(s_{co,m}) + f'^2_Y(s_{co,m})}} \right].$$

Generally speaking, the distance along with the centerline $s_{co,m}$ is not easily obtained, however, as depicted in Figure 3.1, it could be approximated by the actual traveled distance of the obstacle vehicle $s_{co,m}$ in normal driving situations, such that for the mth obstacle, $s_{co,m} \approx s_{o,m}$.

Obviously, U_{obst} is still harmonic based on the properties of the harmonic function, namely, the result of the sum of the harmonic function is still a harmonic function. The safety circle covers the entire obstacle vehicle where its center located in the center of the obstacle.

However, the original obstacle potential (3.24) has shortcomings: the ego vehicle would only start to perform the obstacle avoidance maneuver when near the obstacle, which leads to the path with large curvatures and could be dangerous at high speeds. To guide the host vehicle to avoid the obstacles, we introduce the guiding obstacle potential. As shown in Figure 3.3b, the guiding potential affects the front and rear areas of the obstacle vehicle, namely,

$$U_{gu} = \begin{cases} -k_{gu} \sum_{m=1}^{M} \begin{array}{l} \ln(||\boldsymbol{R}_{t,m} - \boldsymbol{R}_{o,m}|| - r_{sc})(x^*_{t,m} \leq 0) \\ + \ln(||\boldsymbol{R}_{t,m} - \boldsymbol{R}_{o,m}|| - r_{sc})(x^*_{t,m} > 0) \end{array}, & \text{if} ||\boldsymbol{R}_{t,m} - \boldsymbol{R}_{o,m}|| > r_{sc} \\ + \infty, & \text{otherwise}, \end{cases} \quad (3.25)$$

where k_{gu} denotes the stiffness of the guiding potential.

Additionally, as depicted in Figure 3.3, introducing the local obstacle coordinate system $\boldsymbol{O}^*_m$ of the mth obstacle, and the world coordinate of one arbitrary point's position $\boldsymbol{R} = (x, y)$

in the local obstacle coordinate system, O_m^* is denoted by $R_m^* = (x_m^*, y_m^*)$. Similarly, $R_{t,m} = (x_{t,m}, y_{t,m})$ denotes the scaled world coordinates of R, and $R_{t,m}^*$ denotes the position vector of $R_{t,m}$ in the local obstacle coordinate system O_m^*. The rotation relationship in Equation (3.26) holds between R and R_m^*, as well as for $R_{t,m}$ and $R_{t,m}^*$, namely,

$$\begin{cases} R_m^* = \mathcal{R}(R - R_{o,m}) \\ R_{t,m}^* = \mathcal{R}(R_{t,m} - R_{o,m}) \end{cases}, \quad \mathcal{R} = \begin{bmatrix} \cos\varphi_{o,m} & \sin\varphi_{o,m} \\ -\sin\varphi_{o,m} & \cos\varphi_{o,m} \end{bmatrix}, \tag{3.26}$$

where $\varphi_{o,m}$ denotes the heading angle of the mth obstacle, and R_m^* and $R_{t,m}^*$ are connected by the rear scale factor of the guiding obstacle potential of the mth obstacle $\beta_{r,m}$, and front scale factor of the guiding obstacle potential of the mth obstacle $\beta_{f,m}$, namely,

$$\begin{cases} \begin{bmatrix} x_{t,m}^* \\ y_{t,m}^* \end{bmatrix} = \begin{bmatrix} \beta_{r,m} & 0 \\ 0 & 1 \end{bmatrix} \begin{bmatrix} x_m^* \\ y_m^* \end{bmatrix}, & \text{if } x_m^* \leq 0 \\ \begin{bmatrix} x_{t,m}^* \\ y_{t,m}^* \end{bmatrix} = \begin{bmatrix} \beta_{f,m} & 0 \\ 0 & 1 \end{bmatrix} \begin{bmatrix} x_m^* \\ y_m^* \end{bmatrix}, & \text{if } x_m^* > 0 \end{cases}, \quad 0 < \beta_{r,m}, \beta_{f,m} < 1. \tag{3.27}$$

The value of the scale factor $\beta_{r,m}, \beta_{f,m}$ in front and rear of the mth obstacle could be different, such that the potential effect of the obstacle could be more convenient to be manipulated. Set the scale matrix as

$$\Gamma_r = \begin{bmatrix} \beta_{r,m} & 0 \\ 0 & 1 \end{bmatrix}, \quad \Gamma_f = \begin{bmatrix} \beta_{f,m} & 0 \\ 0 & 1 \end{bmatrix}.$$

Noting $x_m^* = \mathcal{R}(R - R_{o,m})e_{ox,m}^*$, $e_{ox,m}^* = (1,0)$ is the unit vector of the local obstacle coordinate system, then Equation (3.27) is rewritten as

$$R_{t,m}^* = \begin{cases} \Gamma_r R_m^* & \text{if } \mathcal{R}(R_{t,m} - R_{o,m})e_{ox,m}^* \leq 0 \\ \Gamma_f R_m^* & \text{if } \mathcal{R}(R_{t,m} - R_{o,m})e_{ox,m}^* > 0 \end{cases}. \tag{3.28}$$

Combining Equation (3.26), the scaled coordinate regarding the mth obstacle is

$$R_{t,m} = \begin{cases} (\mathcal{R}^{-1}\Gamma_r \mathcal{R})(R - R_{o,m}) + R_{o,m} & \text{if } \mathcal{R}(R - R_{o,m})e_{ox,m}^* \leq 0 \\ (\mathcal{R}^{-1}\Gamma_f \mathcal{R})(R - R_{o,m}) + R_{o,m} & \text{if } \mathcal{R}(R - R_{o,m})e_{ox,m}^* > 0. \end{cases} \tag{3.29}$$

Set the total transformation matrix as

$$\begin{cases} \Gamma_r = \mathcal{R}^{-1}\Gamma_r \mathcal{R} \\ \Gamma_f = \mathcal{R}^{-1}\Gamma_f \mathcal{R}. \end{cases}$$

Finally, the expression of the guiding obstacle potential equation (3.25) is rewritten as

$$U_{gu} = \begin{cases} -k_{gu} \sum_{m=1}^{M} \begin{pmatrix} \ln(\|\Gamma_r(R - R_{o,m})\| - r_{sc})([\mathcal{R}(R - R_{o,m})]e_{ox,m}^* \leq 0) \\ + \ln(\|\Gamma_f(R - R_{o,m})\| - r_{sc})([\mathcal{R}(R - R_{o,m})]e_{ox,m}^* > 0) \end{pmatrix}, & C1 \\ + \infty, & \text{otherwise,} \end{cases} \tag{3.30}$$

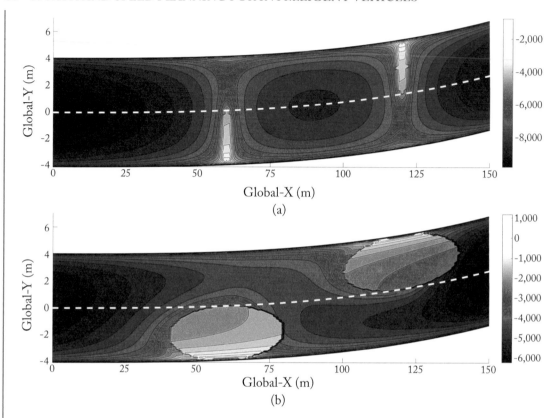

Figure 3.4: Contour plot of the obstacle: (a) The original obstacle potential and (b) The guiding obstacle potential.

where $C1 = $ if $\left((\boldsymbol{\Gamma}_r || \boldsymbol{R}_{t,m} - \boldsymbol{R}_{o,m} || > r_{sc}) \& (\boldsymbol{\Gamma}_f || \boldsymbol{R}_{t,m} - \boldsymbol{R}_{o,m}) > r_{sc} \right)$ represent the condition function.

The influence range of the obstacle would be enlarged through the transformation shown in Equation (3.30). As a consequence, the host vehicle could take obstacle avoidance maneuvers in advance. Figure 3.4 shows the contour distribution of the total potential of two obstacle vehicles lying on the road. Compared with the case with the original obstacle potential, we can observe the potential values with the guiding obstacle potential having been significantly increased in its longitudinal area, which is expected to lead the host vehicle to take obstacle avoidance maneuver in advance.

Finally, the sum of the road boundary potential and the guiding potential is called the external potential (corresponding to the internal potential of the elastic band),

$$U_{ext} = U_{border} + U_{gu}. \tag{3.31}$$

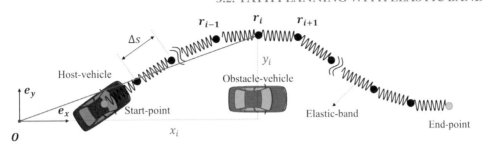

Figure 3.5: Illustration of the elastic band.

The meaning of the external potential is relatively obvious, which reflects the risk of the current driving environment. Since the position and state of the obstacle may change varying with time, thus, it is time dependent.

3.2.2 ELASTIC BANDS

The key idea of the elastic band is to assume the path of obstacle avoidance is represented by virtual elastic bands, which consist of N-pieces of linear springs with $N + 1$ control nodes, as shown in Figure 3.5. The position vector of each node $r_i (i = 1, 2, \ldots, N + 1)$ can be expressed in the world coordinate system as

$$r_i = x_i e_x + y_i e_y, i = i = 1, 2, \ldots, N + 1 \in \mathbb{N}^+. \tag{3.32}$$

Assuming that both ends of the elastic band are fixed, this means that the coordinates of the nodes at the beginning and end are already known. Also, the transformation of those springs satisfy the simple Hooke's law, such that the internal elastic potential at each node of the elastic band U_{int} can be expressed by

$$U_{int,i} = \frac{1}{2}(k_{int}(\|r_{i-1} - r_i\| - l_0)^2 + k_{int}(\|r_{i+1} - r_i\| - l_0)^2), i = 1, 2, \ldots, N \in \mathbb{N}^+, \tag{3.33}$$

where k_{int} denotes the stiffness of the virtual springs, l_0 denotes the natural un-stretched length of the spring, and $\| \cdot \|$ denotes the Euclidean distance, for example,

$$\|r_i - r_{i-1}\| = \sqrt{(x_i - x_{i-1})^2 + (y_i - y_{i-1})^2}. \tag{3.34}$$

Then each node of the elastic-band will be subject to the repulsive force generated by the **external potentials**, which includes the repulsive force from the road border F_{border}, and the guiding force from the obstacle F_{gu} for each node by calculating the negative gradient of the corresponding potential, namely,

$$F_{ext} = F_{border} + F_{gu} = -\nabla U_{ext} = -\nabla U_{border} - \nabla U_{gu}, \tag{3.35}$$

where ∇ denotes the 2D gradient operator. More specifically,

$$\boldsymbol{F}_{border} = -\nabla U_{border} = K\nabla\mathcal{L}, \tag{3.36}$$

where

$$K = \left(\frac{k_{border,l}}{||b-\mathcal{L}||} + \frac{k_{border,r}}{||b+\mathcal{L}||}\right), \ \mathcal{L} = \sqrt{1+\mu^2}(\boldsymbol{R}\cdot\boldsymbol{e_y} - f_Y(s)), \ \mu = \frac{f'_Y(s)}{f'_X(s)}.$$

Also, $\mathcal{F} = \boldsymbol{Z}^{-1}(\boldsymbol{R})$ denotes the corresponding Frenet coordinate of the arbitrary position vector $\boldsymbol{R}$, such that

$$\nabla\mathcal{L} = \frac{\boldsymbol{R}\cdot\boldsymbol{e_y} - f_Y(s)}{\sqrt{1+\mu^2}}\mu\nabla\mu + \sqrt{1+\mu^2}\nabla(\boldsymbol{R}\cdot\boldsymbol{e_y}) - \sqrt{1+\mu^2}f'_Y(s)\nabla s. \tag{3.37}$$

Noting $\nabla(\boldsymbol{A}\cdot\boldsymbol{B}) = \boldsymbol{A}\times(\nabla\times\boldsymbol{B}) + \boldsymbol{B}\times(\nabla\times\boldsymbol{A}) + (\boldsymbol{A}\cdot\nabla)\boldsymbol{B} + (\boldsymbol{B}\cdot\nabla)\boldsymbol{A}$, such that

$$\nabla(\boldsymbol{R}\cdot\boldsymbol{e_y}) = \boldsymbol{R}\times(\nabla\times\boldsymbol{e_y}) + \boldsymbol{e_y}\times(\nabla\times\boldsymbol{R}) + (\boldsymbol{R}\cdot\nabla)\boldsymbol{e_y} + (\boldsymbol{e_y}\cdot\nabla)\boldsymbol{R}. \tag{3.38}$$

Since $\nabla\times\boldsymbol{e_y} = \boldsymbol{0}$, $\nabla\times\boldsymbol{R} = \boldsymbol{0}$, $(\boldsymbol{R}\cdot\nabla)\boldsymbol{e_y} = \boldsymbol{0}$, then

$$\nabla(\boldsymbol{R}\cdot\boldsymbol{e_y}) = (\boldsymbol{e_y}\cdot\nabla)\boldsymbol{R} = \boldsymbol{e_y}. \tag{3.39}$$

Moreover,

$$\begin{cases}\nabla\mu = \rho\nabla s \\ \nabla s = \nabla(\mathcal{F}\boldsymbol{e_x})\end{cases}, \ \rho = \frac{f''_Y(s)f'_X(s) - f'_Y(s)f''_X(s)}{(f'_X(s))^2}. \tag{3.40}$$

Then, Equation (3.37) is simplified as

$$\nabla\mathcal{L} = \left(\frac{\boldsymbol{R}\cdot\boldsymbol{e_y} - f_Y(s)}{\sqrt{1+\mu^2}}\mu\rho - \sqrt{1+\mu^2}f'_Y(s)\right)\nabla(\mathcal{F}\cdot\boldsymbol{e_x}) + \sqrt{1+\mu^2}\boldsymbol{e_y}. \tag{3.41}$$

Again, we have,

$$\nabla(\mathcal{F}\cdot\boldsymbol{e_x}) = \mathcal{F}\times(\nabla\times\boldsymbol{e_x}) + \boldsymbol{e_x}\times(\nabla\times\mathcal{F}) + (\mathcal{F}\cdot\nabla)\boldsymbol{e_x} + (\boldsymbol{e_x}\cdot\nabla)\mathcal{F}, \tag{3.42}$$

and noting that $\nabla\times\boldsymbol{e_x} = \boldsymbol{0}, (\mathcal{F}\cdot\nabla)\boldsymbol{e_x}) = \boldsymbol{0}, \boldsymbol{e_x}\times(\nabla\times\mathcal{F}) = \nabla_\mathcal{F}(\boldsymbol{e_x}\cdot\mathcal{F}) - (\boldsymbol{e_x}\cdot\nabla)\mathcal{F}$. Therefore,

$$\nabla(\mathcal{F}\cdot\boldsymbol{e_x}) = \nabla_\mathcal{F}(\boldsymbol{e_x}\cdot\mathcal{F}), \tag{3.43}$$

where the notation $\nabla_\mathcal{F}$ means that the subscripted gradient operates only on the $\mathcal{F}$. As described in Equation (3.6),

$$\nabla_\mathcal{F}(\boldsymbol{e_x}\cdot\mathcal{F}) = \left(\boldsymbol{e_x}\cdot\frac{\partial\mathcal{F}}{\partial x}, \boldsymbol{e_x}\cdot\frac{\partial\mathcal{F}}{\partial y}\right) = \left(\frac{\partial z_1^{-1}(\boldsymbol{R})}{\partial x}, \frac{\partial z_1^{-1}(\boldsymbol{R})}{\partial y}\right).$$

According to Equation (3.4), the derivative of the first component of the Frenet coordinate respecting to the world coordinate is calculated by

$$
\begin{cases}
\dfrac{\partial x}{\partial s} = f_X'(s) - \dfrac{L f_Y''(s)\sqrt{f_X'^2(s) + f_Y'^2(s)} - L f_Y'(s)(f_X'(s) f_X''(s) + f_Y'(s) f_Y''(s))}{(f_X'^2 + f_Y'^2(s))^{\frac{3}{2}}} \\[4mm]
\dfrac{\partial y}{\partial s} = f_Y'(s) + \dfrac{L f_X''(s)\sqrt{f_X'^2(s) + f_Y'^2(s)} + L f_X'(s)(f_X'(s) f_X''(s) + f_Y'(s) f_Y''(s))}{(f_X'^2 + f_Y'^2(s))^{\frac{3}{2}}}.
\end{cases}
$$

Therefore,

$$
\frac{\partial}{\partial x} z_1^{-1}(\boldsymbol{R}) = 1 / \frac{\partial x}{\partial s}, \frac{\partial}{\partial y} z_1^{-1}(\boldsymbol{R}) = 1 / \frac{\partial y}{\partial s}.
$$

Then,

$$
\begin{aligned}
\nabla \mathcal{L} =\ & \left(\frac{\boldsymbol{R} \cdot \boldsymbol{e_y} - f_Y(s)}{\sqrt{1 + \mu^2}} \mu \rho - \sqrt{1 + \mu^2}\, f_Y'(s) \right) \frac{\partial}{\partial x} z_1^{-1}(\boldsymbol{R}) \boldsymbol{e_x} \\
& + \left(\left(\frac{\boldsymbol{R} \cdot \boldsymbol{e_y} - f_Y(s)}{\sqrt{1 + \mu^2}} \mu \rho - \sqrt{1 + \mu^2}\, f_Y'(s) \right) \frac{\partial}{\partial y} z_1^{-1}(\boldsymbol{R}) + \sqrt{1 + \mu^2} \right) \boldsymbol{e_y}.
\end{aligned}
\tag{3.44}
$$

Finally, rewrite the analytical expression of the virtual border potential force at each node as

$$
\begin{aligned}
\boldsymbol{F}_{border,i} &= \left(K_i \xi_i \frac{\partial}{\partial x} z_1^{-1}(\boldsymbol{r}_i) \right) \boldsymbol{e_x} + \left(K_i \xi_i \frac{\partial}{\partial y} z_1^{-1}(\boldsymbol{r}_i) + K_i \sqrt{1 + \mu_i^2} \right) \boldsymbol{e_x} \\
K_i &= \left(\frac{k_{border,l}}{\|b - \mathcal{L}_\rangle\|} + \frac{k_{border,r}}{\|b + \mathcal{L}_\rangle\|} \right), \ \mathcal{L}_\rangle = \sqrt{1 + \mu_i^2}(r_i \boldsymbol{e_y} - f_Y(s)) \\
\xi_i &= \frac{\boldsymbol{r}_i \cdot \boldsymbol{e_y} - f_Y(s)}{\sqrt{1 + \mu_i^2}} \mu_i \rho_i - \sqrt{1 + \mu_i^2}\, f_Y'(s), \mu = \frac{f_Y'(s)}{f_X'(s)} \\
\rho_i &= \frac{f_Y''(s_i) f_X'(s_i) - f_Y'(s_i) f_X''(s_i)}{(f_X'(s_i))^2}, \ s_i = \boldsymbol{Z}^{-1}(\boldsymbol{r}_i) \cdot \boldsymbol{e_x}, \ L_i = \boldsymbol{Z}^{-1}(\boldsymbol{r}_i) \cdot \boldsymbol{e_x}.
\end{aligned}
\tag{3.45}
$$

Besides,

$$
\frac{\partial}{\partial x} z_1^{-1}(\boldsymbol{r}_i) = \frac{1}{f_X'(s_i) - \dfrac{L_i f_Y''(s_i)\sqrt{f_X'^2(s_i) + f_Y'^2(s_i)} - L_i f_Y'(s_i)(f_X'(s_i) f_X''(s_i) + f_Y'(s_i) f_Y''(s_i))}{(f_X'^2(s) + f_Y'^2(s))^{3/2}}}
$$

$$
\frac{\partial}{\partial y} z_1^{-1}(\boldsymbol{r}_i) = \frac{1}{f_Y'(s_i) + \dfrac{L_i f_X''(s_i)\sqrt{f_X'^2(s_i) + f_Y'^2(s_i)} + L_i f_X'(s_i)(f_X'(s_i) f_X''(s_i) + f_Y'(s_i) f_Y''(s_i))}{(f_X'^2(s) + f_Y'^2(s))^{3/2}}}
$$

As for the guiding obstacle potential force, at first, let r denotes any 2D position vector in Euclidean space; then,

$$\nabla ||r|| = \nabla \sqrt{r \cdot r} = \frac{\nabla \sqrt{r \cdot r}}{2||r||}, \ \nabla(r \cdot r) = 2r \times (\nabla \times r) + 2(r \cdot \nabla)r,$$

since $\nabla \times r = 0, (r \cdot \nabla)r = r$, which yields

$$\nabla(r \cdot r) = 2r, \nabla ||r|| = \frac{r}{||r||}. \tag{3.46}$$

That means, for the position vector from an arbitrary location on the road to the obstacle vehicle center, Equation (3.47) holds:

$$\nabla ||R - R_{o,m}|| = \frac{R - R_{o,m}}{||R - R_{o,m}||}. \tag{3.47}$$

So, the virtual potential force generated by the guiding obstacle potential at each node is calculated by

$$F_{gu,i} =$$
$$\begin{cases} k_{gu} \sum_{m=1}^{M} \left(\dfrac{\Gamma_r}{||\Gamma_r(r_i - R_{o,m})|| - r_{sc}} \cdot \dfrac{r_i - R_{o,m}}{||r_i - R_{o,m}||}([\mathcal{R}(R - R_{o,m})]e^*_{ox,m} \leq 0) \\ \qquad + \dfrac{\Gamma_f}{||\Gamma_f(r_i - R_{o,m})|| - r_{sc}} \cdot \dfrac{r_i - R_{o,m}}{||r_i - R_{o,m}||}([\mathcal{R}(R - R_{o,m})]e^*_{ox,m} \leq 0) \right), \ C2 \\ + \infty, \ \text{otherwise,} \end{cases}$$

$$\tag{3.48}$$

where $C2 = \text{if} \left(\Gamma_r ||r_i, tm - R_{o,m}|| > r_{sc} \& \Gamma_f ||r_i, tm - R_{o,m}|| > r_{sc} \right)$ is the condition function. Finally, the external force of each node is calculated by

$$F_{ext,i} = F_{border,i} + F_{gu,i}. \tag{3.49}$$

Similarly, the internal force of each node is calculated by the negative gradient of the internal potential between adjacent nodes, which is the sum of the elastic force $F_{int,l}$ on the left side and the elastic force $F_{int,r}$ on the right side; moreover, Equation (3.50) holds for the position vector between adjacent nodes,

$$\begin{cases} \nabla ||r_{i+1} - r_i = \dfrac{r_{i+1} - r_i}{||r_{i+1} - r_i||} \\ \nabla ||r_{i-1} - r_i = \dfrac{r_{i-1} - r_i}{||r_{i-1} - r_i||}. \end{cases} \tag{3.50}$$

Therefore, the virtual internal force is evaluated by

$$F_{int,i} = F_{int,l} + F_{int,r} = -\nabla U_{int,i} = K_r r_i - (K_r + K_l)r_i + K_l r_{i-1}, \tag{3.51}$$

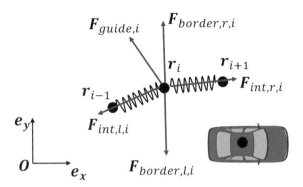

Figure 3.6: Illustration of the virtual forces acting on the node.

where

$$K_r = -\frac{k_{int}(||r_{i+1} - r_i|| - l_0)}{||r_{i+1} - r_i||}, K_l = -\frac{k_{int}(||r_{i-1} - r_i|| - l_0)}{||r_{i-1} - r_i||}.$$

According to the force equilibrium state at each node, the position of the node could be obtained by solving the following equation sets, namely,

$$\{F_{ext,i} + F_{int,i} = 0, i = 2, \ldots, N \in \mathbb{N}^+\}. \tag{3.52}$$

Decomposing Equation (3.52) into e_x and e_y direction to obtain a $2(N - 1)$-size of equation sets. However, for the simplicity, restraining the horizontal gap between adjacent nodes to be evenly distributed by Δx, such that the horizontal component of the world coordinate of the elastic-bands r_i is calculated by

$$r_i \cdot e_x = x_i = i \cdot \Delta x, i = 1, 2, \ldots, N \in \mathbb{N}^+. \tag{3.53}$$

Therefore, we only need $(N - 1)$ to obtain the vertical displacement of all elastic band nodes in world coordinates, such that (3.52) can be further simplified as

$$\{F_{ext,y,i} + F_{int,y,i} = 0, i = 2, \ldots, N \in \mathbb{N}^+\}, \tag{3.54}$$

where $F_{ext,i}$ and $F_{int,i}$ denote the vertical virtual external force and the vertical virtual internal force acting on each node.

3.2.3 SOLUTION OF THE ELASTIC BANDS

From the above derivations, it is concluded that the position of the nodes could be obtained by solving the nonlinear equation sets (3.54). Generally speaking, most of the nonlinear equation sets can be solved by numerical methods such as the Newton–Raphson method, Broyden's method, etc., and the Newton–Raphson method and its variations are the most representative numerical solutions for solving nonlinear systems. The Newton–Raphson method is a

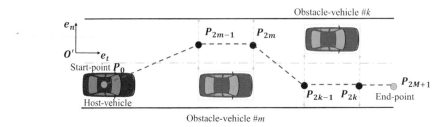

Figure 3.7: Illustration of the initial solution generation for elastic bands.

root-finding algorithm producing successively better approximations to the roots of real-valued functions, which requires the function to be solved is continuously differentiable, and its first derivative shall be continuous as well. Moreover, the convergence speed is acceptable when a sufficiently good initial value is afforded.

To solve (3.54), applying a linear Taylor expansion to the left side of the equation while ignoring those high-order terms, it can be approximated as follows:

$$F(z) \approx F(z^{(k)}) + J(z^{(k)})(z - z^{(k)}) = 0, \dim z = N - 1, k = 0, 1, 2, \ldots, \in \mathbb{N}, \quad (3.55)$$

where $F = [F_{totoal,1}, F_{totoal,2}, \ldots, F_{totoal,N-1}]$, J denotes the Jacobian matrix of F, k denotes the iteration number, and z denotes the solution variable.

Set the solution of Equation (3.55) as $z^{(k+1)}$, yielding,

$$z^{(k+1)} = z^{(k)} - J^{-1}(z^{(k)})F(z^{(k)}). \quad (3.56)$$

Repeated the above processing until the solution error $\Delta z^{(k)}$ lies within the predefined small positive number ε, namely,

$$\Delta z^{(k)} = z^{(k+1)} - z^{(k)} \leq \varepsilon, \varepsilon > 0.$$

The initial solution is denoted by $z^{(0)} = [z_1^{(0)}, z_2^{(0)}, \ldots, z_{N-1}^{(0)}]$ shall be given at the very beginning of the process. In our work, a simple geometric method in the Frenet coordinate system can be utilized for generating initial solutions. As shown in Figure 3.7, key midpoints $P_{2m}, P_{2m-1}, (1 \leq m \leq M, m \in \mathbb{N}^+)$ in Frenet coordinates at both ends for each obstacle was located (total number of obstacles M), where they lie between the furthest road boundary and the boundary of the obstacle. Then along with the start point marked as P_0 and the target point marked as P_{2M+1} were used for the interpolation, if the longitudinal distance of the Frenet coordinate for each node is equally distributed, such that the lateral distance of the Frenet coordinate for all nodes could be obtained accordingly. Finally, they will be transforming back to the world coordinates to generate the initial solution sets $z_{(0)}$ for all elastic-band nodes.

Furthermore, since the hazardous potential of the environment is time-depended, the path needs to be updated with a certain fresh frequency to deal with moving obstacles, then the last planned-path can be used as the initial value of the next stage of the path-planning procedure.

Table 3.1: Parameter settings of the elastic band

Symbol	Value	Description
$k_{border,l}$	1,000 N/m	Stiffness of the left border potential
$k_{border,r}$	1,000 N/m	Stiffness of the right border potential
k_{int}	1,000 N/m	Stiffness of the elastic band
k_{obst}	500 N/m	Stiffness of the obstacle potential
k_{gu}	500 N/m	Stiffness of the guiding obstacle potential
r_{sc}	2 m	The radius of the safety circle
βr	0.20	The rear scale factor of the guiding obstacle potential
βf	0.10	The front scale factor of the guiding obstacle potential
l_0	8.0 m	Un-stretched length of the elastic band
Δx	10 m	The longitudinal gap of elastic band nodes
ε	1e-6	Newton-Raphson convergence error

3.2.4 STUDY CASES

The simulation scenario is set as follows. Assume that the host vehicle detects obstacles in front and begins to take obstacle avoidance maneuvers, where the initial position of the host vehicle is on the center of the right lane, while the starting position of the static obstacle #1 is 60 m away from the lane in (right lane) with the host vehicle, and the static obstacle #2 is 120 m away from the center of the adjacent lane (left lane). Moreover, the width of the lane is 4 m and the goal position is the center of the right lane that is 150 m away from the host vehicle. The planned path would be generated by the elastic band with the guiding obstacle potential and one with the original for the comparative purpose; the algorithm-related parameter settings are shown in Table 3.1.

The planned path for obstacle avoidance is shown in Figure 3.8. The green dashed line indicates the planned path by the guiding obstacle potential is introduced, while the blue dotted line indicates the planned path with the original elastic band method. The simulation results show that the planned path with the guiding obstacle potential improves significantly, which also tends to be smoother and more natural. Moreover, due to that, the impact of the obstacle potential on the host vehicle becomes larger, the host vehicle will enter into the adjacent lane to avoid obstacles #2 earlier and keep a larger lateral distance to the obstacle as well, thereby making the host vehicle safer in the process of avoiding obstacles.

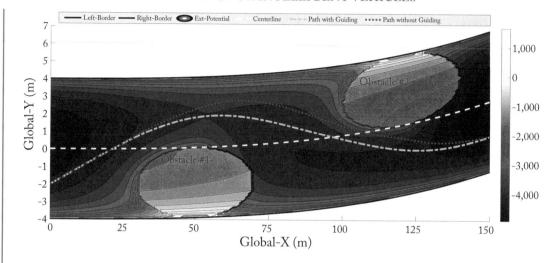

Figure 3.8: The path planning results based on the hazard potential and elastic bands for obstacle avoidance.

3.3 PATH PLANNING WITH HARMONIC POTENTIALS FOR OBSTACLE AVOIDANCE

The path planning with elastic bands introduced in the previous section discretizes the path into a series of control points, then obtain those positions by solving a nonlinear equation set. In this way, the connections of the generated path through control points are smooth without large sudden changes in curvatures. However, the shortcomings of this method are also obvious. We have to solve a system of nonlinear equations; generally speaking, it will consume more computing resources, especially when there are a large number of control points to be solved. Such that it cannot be met with real-time application requirements. Therefore, we need to come up with ideas to speed up the solution finding. Considering that most obstacle avoidance on the road can be realized by a lane change maneuver for the normal driving, we will focus on the path planning for the lane change in this section. Once again, we shall combine the guiding obstacle potential introduced in the previous section; however, we will plan it on the Frenet coordinate space to simplify the representation of the potentials of the obstacle and road borders. Meanwhile, use the fastest gradient descent method to obtain the path, instead of the elastic band, to improve the calculation efficiency of the path planning with artificial potentials. Finally, transform the path to the world.

Since no virtual elastic bands are connecting those nodes, we need to introduce an additional potential, which is the **attractive goal potential** so-called at the target point, such that it can guide the host vehicle toward the target position. The attractive goal potential U_{goal} can be

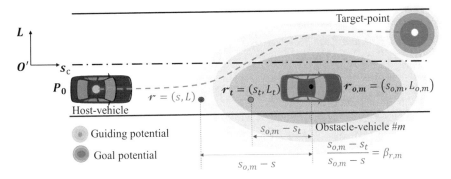

Figure 3.9: The illustration of the hazard potential construction on Frenet coordinate space.

designed as follows:

$$U_{goal} = k_{goal}\|\boldsymbol{F}_{goal} - \boldsymbol{F}\|, \tag{3.57}$$

where $k_{goal} > 0$ denotes the stiffness of the target potential, $\boldsymbol{F}_{goal} = (s_{goal}, L_{goal})$ denotes the Frenet coordinates of the target point's position, and $\boldsymbol{F} = (s, L)$ denotes the Frenet coordinate of an arbitrary point's position.

Besides, the road border potential and the guiding obstacle potential are all described in this Frenet coordinate system with logarithmic functions, namely,

$$U_{border} = -k_{border,l} \ln|b - L| - k_{border,r} \ln|L + b| \tag{3.58}$$

$$U_{guide} = \begin{cases} -k_{guide} \sum_{m=1}^{M} \left(\begin{array}{l} \ln(\|\boldsymbol{\Gamma_r}(\boldsymbol{F_m} - \boldsymbol{F_{o,m}})\| - r_{sc}) \cdot (s_{t,m} \leq s_{o,m}) \\ + \ln(\|\boldsymbol{\Gamma_f}(\boldsymbol{F_m} - \boldsymbol{F_{o,m}})\| - r_{sc}) \cdot (s_{t,m} > s_{o,m}) \end{array} \right), C3 \\ +\infty, \text{ otherwise,} \end{cases} \tag{3.59}$$

where $k_{border,l}$ and $k_{border,r}$ denote the stiffness of the left and right road boundary potential, k_{guide} denotes the stiffness of the guiding obstacle potential, b denotes the lane width, r_{sc} denotes the radius of the safety circle. $\boldsymbol{F}_{o,m} = (s_{o,m}, L_{o,m})$ denotes the center position Frenet coordinate of the mth obstacle, $\boldsymbol{F}_{t,m} = (s_{t,m}, L_{t,m})$ denotes the scaled Frenet coordinates for the rear (front) area of the obstacle, which is related to the rear (front) scale factor of the guiding potential of the mth obstacle $\beta_{(r,m)}\beta_{(f,m)}$ regarding the rear(front) scale matrix $\boldsymbol{\Gamma_r}(\boldsymbol{\Gamma_f})$, also

$$\boldsymbol{\Gamma_r} = \begin{bmatrix} \beta_{(r,m)} & 0 \\ 0 & 1 \end{bmatrix}, \boldsymbol{\Gamma_f} = \begin{bmatrix} \beta_{(f,m)} & 0 \\ 0 & 1 \end{bmatrix}.$$

Finally, the external potential U_{ext} is calculated by

$$U_{ext} = U_{border} + U_{guide}, \tag{3.60}$$

Algorithm 3.2 Pseudo-code of path planning based on the maximum gradient descent method.

Initialization: $\boldsymbol{F_{x,0}} = S_{start}$, $\boldsymbol{F_{y,0}} = L_{start}$, $k = 1$
Output: The planned path in Frenet coordinates $(\boldsymbol{F_s}, \boldsymbol{F_L})$.

1: **while** $\sqrt{(\boldsymbol{F_{s,k-1}} - s_{goal})^2 + (\boldsymbol{F_{L,k-1}} - L_{goal})^2} \leq \varepsilon$ **do**

2: $\quad v_{s,k} = -\dfrac{\partial U_{totoal}}{\partial s}\Big|_{s=\boldsymbol{F_{s,k-1}},L=\boldsymbol{F_{L,k-1}}}$, $v_{L,k} = -\dfrac{\partial U_{totoal}}{\partial L}\Big|_{s=\boldsymbol{F_{s,k-1}},L=\boldsymbol{F_{L,k-1}}}$

3: $\quad v_{\bar{s},k} = \dfrac{v_{s,k}}{\sqrt{v_{s,k}^2 + v_{L,k}^2}}$, $v_{\bar{L},k} = \dfrac{v_{L,k}}{\sqrt{v_{s,k}^2 + v_{L,k}^2}}$

4: $\quad \boldsymbol{F_{s,k+1}} = \boldsymbol{F_{s,k}} + \mathrm{d}s \cdot v_{\bar{s},k}$, $\boldsymbol{F_{L,k+1}} = \boldsymbol{F_{L,k}} + \mathrm{d}s \cdot v_{\bar{L},k}$

5: $\quad k = k + 1$

6: **end while**

---

and the total potential is defined as

$$U_{totoal} = U_{ext} + U_{goal}. \tag{3.61}$$

If the host vehicle is regarded as a particle moving under the effect of the repulsive external potential and the attractive goal potential, its position $\boldsymbol{F}$ can be governed by the gradient system, as shown in Equation (3.62); that is,

$$\frac{\mathrm{d}\boldsymbol{F}}{\mathrm{d}t} = -\nabla U_{total}. \tag{3.62}$$

The pseudo-code process of path solution is shown in Table 3.2. The principle of the pseudo-code is that if the distance between the current position and the target point is greater than a predefined threshold ε, then the unit vector $\langle v_{\bar{s},k}, v_{\bar{L},k} \rangle$ of the point is calculated based on the current position with a step-size ε. Finally, the position of the object at the next iteration can be approximated by (3.63),

$$\begin{cases} \boldsymbol{F_{s,k+1}} = \boldsymbol{F_{s,k}} + \mathrm{d}s \cdot v_{\bar{s},k} \\ \boldsymbol{F_{L,k+1}} = \boldsymbol{F_{L,k}} + \mathrm{d}s \cdot v_{\bar{L},k}. \end{cases} \tag{3.63}$$

The process in Equation (3.63) avoids solving complicated nonlinear equations of the elastic band method, which improves the speed of solution-finding. Finally, the path points in the Frenet coordinate space would transfer back to the Cartesian coordinate space via the relationship, as shown in Equation (3.3).

According to the simulation experience, the stiffness of the attractive potential at the goal point, as well as the scale factor of the guiding obstacle potential, have greater impacts on the

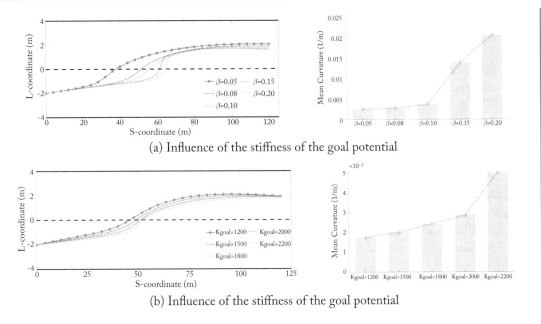

(a) Influence of the stiffness of the goal potential

(b) Influence of the stiffness of the goal potential

Figure 3.10: Influencing factors study.

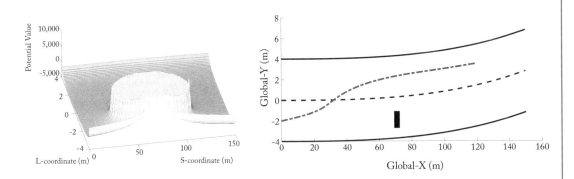

Figure 3.11: Path planning for obstacle avoidance.

path generation. Thus, we will focus on analyzing these two parameters' influences on the path. The test scenario is described as follows, assuming that one obstacle lies in the center of the right lane which is 70 m away from the host vehicle and keeps the lateral distance to the centerline is -2 m (right), namely, the Frenet coordinate of the obstacle center is (70 m, -2 m). Moreover, the road is curved where the centerline position (x_c, y_c) can be represented by a cubic function $y_c = (x_c^3)/(1.2 10^6)$, and the lane width is 4 m. The target position in the Frenet coordinate system is (120 m, 2 m).

As shown in Figure 3.10a, the smaller of the scale factor of the guiding potential is, when the value of the scale factor of the guiding potential β (set $\beta_r = \beta_f = \beta$ in this case) varies from 0.20, 0.15, 0.10, 0.08 to 0.05, respectively, the larger of the traveled distance of the obstacle when started to enter the adjacent lane, which varied approximately from 40 to 65 m. Moreover, the path also tends to be smoother where the mean curvature of the path is smaller, as shown in Figure 3.10a. Therefore, the value of the scale factor of the guiding potential β should not be large, and the recommended value is 0.05 in this case.

As for the stiffness of the goal potential, as shown in Figure 3.10b, when the value of k_{goal} ranges from 1200, 1500, 1800, 2000, to 2200, a larger k_{goal} tends to lead the path with larger curvatures. Therefore, to obtain a more natural lane-change shape, we recommend that the parameter value of k_{goal} should not be too large as well, and the recommended value, in this case, is 1200. Finally, with the combination of the scale factor of the guiding potential $\beta = 0.05$ and the stiffness of the goal potential $k_{goal} = 1200$, the path for obstacle avoidance on a cured road is shown in Figure 3.11.

3.4   OPTIMAL PATH PLANNING WITH NATURAL CUBIC SPLINES

Although a feasible path could be generated based on artificial potential methods, still, we cannot ensure the collision-free path to be optimal for its length and curvature. Therefore, we aim to develop an optimal path-planning method in this section, which is based on the natural cubic spline and convex optimization for any shapes of roads.

Recalling the content in Section 3.1, the global coordinates of any arbitrary point on the road can be described by a pair of Frenet coordinate (s, L), namely, the longitudinal distance s which is the length of the traveled path along the road centerline, and the lateral distance L to the centerline from the point. The relationship of one point between the Cartesian coordinate and the Frenet coordinate is described by the following equation:

$$\boldsymbol{R}(s) = (x_c(s) + L\chi(s))\boldsymbol{e_x} + (y_c(s) + L\sigma(s))\boldsymbol{e_y}, \tag{3.64}$$

where x_c, y_c denote the Cartesian coordinates of the centerline, which is a function of the longitudinal distance s of the Frenet coordinate, and

$$\chi(s) = -\frac{f'_Y(s)}{\sqrt{f'^2_X(s) + f'^2_Y(s)}}, \quad \sigma(s) = -\frac{f'_X(s)}{\sqrt{f'^2_X(s) + f'^2_Y(s)}}. \tag{3.65}$$

Assume that the planned path is composed of $n + 1$ points, and the distances between the nodes are kept evenly distributed. In this way, the node positions can be written in the following matrix form:

$$\begin{cases} x = x_c + \mathbf{diag}[\chi_i]L \\ y = y_c + \mathbf{diag}[\sigma_i]L \end{cases}, i = 1, 2, \ldots, n + 1, \tag{3.66}$$

where $\boldsymbol{L} = [L_0, L_1, \ldots, L_n]$ represents the lateral distance of each point to the centerline, which is regarded as the control variable of the planned path.

To generate a suitable path, let us consider the following indicators: (1) the shortest driving path, which is conducive to the time-saving and fuel consumption reduction; (2) the minimal curvature, whose curvature of the generated path is designed to avoid the occurrence of dangerous conditions due to large curvatures; and (3) the minimal path heading error, which ensures the planned path consistent with the overall heading of the road. We will construct related quadratic terms for those indicators in detail below.

3.4.1   THE TRAVELED DISTANCE

The difference between two adjacent control points is

$$
\begin{cases} \Delta x_i = (x_{c,i+1} - x_{c,i}) + (\chi_{i+1} L_{i+1} - \chi_i L_i) = \Delta x_{ci} + \boldsymbol{\alpha_i} \bar{\boldsymbol{L_i}} \\ \Delta y_i = (y_{c,i+1} - y_{c,i}) + (\sigma_{i+1} L_{i+1} - \sigma_i L_i) = \Delta y_{ci} + \boldsymbol{\beta_i} \bar{\boldsymbol{L_i}}, \end{cases} \tag{3.67}
$$

where

$$
\boldsymbol{\alpha_i} = [-\chi_i \ \chi_{i+1}], \boldsymbol{\beta_i} = [-\sigma_i \ \sigma_{i+1}], \bar{\boldsymbol{L_i}} = [L_i \ L_{i+1}]^T.
$$

Then the quadratic term S^2 related to the whole curve,

$$
S^2 = \sum_{i=1}^n ((\Delta x_i)^2 + (\Delta y_i)^2) = \sum_{i=1}^n \left(\bar{\boldsymbol{L_i}}^T (\boldsymbol{\alpha_i}^T \boldsymbol{\alpha_i} + \boldsymbol{\beta_i}^T \boldsymbol{\beta_i}) \bar{\boldsymbol{L_i}} + 2 \Delta_i^T \boldsymbol{\gamma_i} \bar{\boldsymbol{L_i}} + \Delta_i^T \Delta_i \right), \tag{3.68}
$$

where $\Delta_i = [\Delta x_{ci} \ \Delta y_{ci}]^T$, $\boldsymbol{\gamma_i} = [\boldsymbol{\alpha_i} \ \boldsymbol{\beta_i}]^T$. Equation (3.68) can be further expressed by

$$
S^2 = \bar{\boldsymbol{L_i}}^T \text{diag}[\boldsymbol{\alpha_i}^T \boldsymbol{\alpha_i} + \boldsymbol{\beta_i}^T \boldsymbol{\beta_i}] \bar{\boldsymbol{L_i}} + 2 \Delta_i^T \text{diag}[\boldsymbol{\gamma_i}] \bar{\boldsymbol{L_i}} + \Delta_i^T \Delta_i. \tag{3.69}
$$

Assuming $\bar{\boldsymbol{L_i}} = \boldsymbol{g} \boldsymbol{L}$, where $\boldsymbol{g}$ is a constant matrix, the final form of Equation (3.69) is

$$
S^2 = \boldsymbol{L}^T \mathbf{K} \boldsymbol{L} + \mathbf{M} \boldsymbol{L} + \text{cost}, \tag{3.70}
$$

where the cost is a constant term that is independent of $\boldsymbol{L}$, and thus it can be omitted:

$$
\begin{cases} \mathbf{K} = \boldsymbol{g}^T \text{diag}[\boldsymbol{\alpha_i}^T \boldsymbol{\alpha_i} + \boldsymbol{\beta_i}^T \boldsymbol{\beta_i}] \boldsymbol{g} \\ \mathbf{M} = 2 \Delta_i^T \text{diag}[\boldsymbol{\gamma_i}] \boldsymbol{g}. \end{cases} \tag{3.71}
$$

3.4.2   THE CURVATURE

To obtain a smooth path, it is assumed that the path is connected through the control points by a natural cubic spline, that is

$$
\begin{cases} x_{ri}(s^*) = a_{0i} + a_{1i} s^* + a_{2i} s^{*2} + a_{3i} s^{*3} \\ y_{ri}(s^*) = b_{0i} + b_{1i} s^* + b_{2i} s^{*2} + b_{3i} s^{*3}, \end{cases} \tag{3.72}
$$

where $s^* = \dfrac{s - s_{i0}}{\Delta s} \in [0, 1]$, and the path curvature square term can be expressed according to the definition of the curvature:

$$K_i^2(s) = \left(\frac{\mathrm{d}s^*}{\mathrm{d}s}\right)^4 \left[\left(\frac{\mathrm{d}^2 x_{ri}(s^*)}{\mathrm{d}s^{*2}}\right)^2 + \left(\frac{\mathrm{d}^2 y_{ri}(s^*)}{\mathrm{d}s^{*2}}\right)^2\right]. \tag{3.73}$$

Ignore the multiplier factors. The sum of the squared curvature of the entire path K^2 is

$$K^2 = \sum_{i=1}^{n} \left[\left(\frac{\mathrm{d}^2 x_{ri}(s^*)}{\mathrm{d}s^{*2}}\right)^2 + \left(\frac{\mathrm{d}^2 y_{ri}(s^*)}{\mathrm{d}s^{*2}}\right)^2\right]. \tag{3.74}$$

According to the conclusions obtained in the previous Section 3.1, the second derivative of the natural cubic spline curve evaluated at the endpoint can be expressed as a linear form regarding the control point, such that

$$\left.\frac{\mathrm{d}^2 \boldsymbol{x}(s^*)}{\mathrm{d}s^{*2}}\right|_{s^*=0} = \mathbf{H}\boldsymbol{x} = \mathbf{H}\boldsymbol{x}_c + \mathbf{H}\mathrm{diag}[\chi_i]\boldsymbol{L} \tag{3.75}$$

$$\left\|\frac{\mathrm{d}^2 \boldsymbol{x}(s^*)}{\mathrm{d}s^{*2}}\right\|_{s^*=0} = \boldsymbol{x}_c^T(\mathbf{H}^T\mathbf{H})\boldsymbol{x}_c + \boldsymbol{L}^T(\mathbf{H}^T\mathbf{H})\mathrm{diag}[\chi_i]\boldsymbol{L} + 2(\boldsymbol{x}_c^T\mathbf{H}^T\mathbf{H})\mathrm{diag}[\chi_i]\boldsymbol{L}. \tag{3.76}$$

Similarly,

$$\left\|\frac{\mathrm{d}^2 \boldsymbol{y}(s^*)}{\mathrm{d}s^{*2}}\right\|_{s^*=0} = \boldsymbol{y}_c^T(\mathbf{H}^T\mathbf{H})\boldsymbol{y}_c + \boldsymbol{L}^T(\mathbf{H}^T\mathbf{H})\mathrm{diag}[\sigma_i]\boldsymbol{L} + 2(\boldsymbol{y}_c^T\mathbf{H}^T\mathbf{H})\mathrm{diag}[\sigma_i]\boldsymbol{L}. \tag{3.77}$$

Notice that

$$\mathrm{diag}[\chi_i^2] + \mathrm{diag}[\sigma_i^2] = \mathrm{diag}[\mathbf{1}].$$

Then K^2 can be also expressed as a quadratic term related to the control points $\boldsymbol{L}$, namely,

$$K^2 = \boldsymbol{L}^T \mathbf{N}\boldsymbol{L} + \mathbf{T}\boldsymbol{L} + \mathrm{cost}. \tag{3.78}$$

Again, the cost is a constant term independent to $\boldsymbol{L}$, $\mathbf{N}$, and $\mathbf{T}$ are calculated by

$$\begin{cases} \mathbf{N} = \mathbf{H}^T\mathbf{H} \\ \mathbf{T} = 2(\boldsymbol{x}_c^T\mathbf{H}^T\mathbf{H})\mathrm{diag}[\chi_i] + 2(\boldsymbol{y}_c^T\mathbf{H}^T\mathbf{H})\mathrm{diag}[\sigma_i]. \end{cases} \tag{3.79}$$

3.4.3   PATH HEADING

The heading error between the planned path and the centerline is expected to be small to avoid large deviations from the road heading. As depicted in Figure 3.12, the heading error $\Delta\vartheta_i$ between the adjacent nodes i and $i - 1$ is approximated by

$$\Delta\vartheta_i = \arctan \frac{L_i - L_{i-1}}{\Delta s_x} \approx \frac{L_i - L_{i-1}}{\Delta s}, \tag{3.80}$$

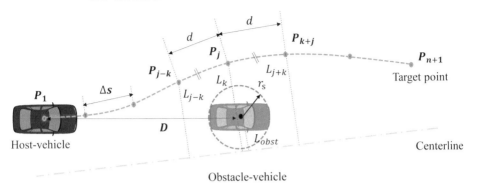

Figure 3.12: Illustration of the elastic band.

where Δs_x denotes the longitudinal distance gap between the adjacent nodes, while Δs denotes the corresponding longitudinal distance gap respecting the road centerline. Then a quadratic term P^2 regarding the heading error can be constructed as follows:

$$P^2 = \sum_{i=1}^{m} \Delta \vartheta_i^2 = \frac{1}{\Delta s^2} \boldsymbol{L}^T (\boldsymbol{w_1} - \boldsymbol{w_2})^T (\boldsymbol{w_1} - \boldsymbol{w_2}) \boldsymbol{L} = \boldsymbol{L}^T \mathbf{Q} \boldsymbol{L}, \tag{3.81}$$

where $\mathbf{Q} = (\boldsymbol{w_1} - \boldsymbol{w_2})^T (\boldsymbol{w_1} - \boldsymbol{w_2})/\Delta s^2$, $\boldsymbol{w_1}, \boldsymbol{w_2}$ denotes constant matrix which satisfies

$$\begin{aligned} \boldsymbol{w_1}\boldsymbol{L} &= [L_1, L_2, \ldots, L_n] \\ \boldsymbol{w_2}\boldsymbol{L} &= [L_0, L_1, \ldots, L_{n-1}]^T. \end{aligned} \tag{3.82}$$

3.4.4   PATH-PLANNING SOLUTION FOR SPECIFIC MANEUVERS

Free Driving

The only restriction in free driving is to keep the vehicle within the road boundaries. Combining all indicators listed in Sections 3.4.1–3.4.3, the optimal path planning for free driving is

$$\begin{aligned} \min \ &\Sigma^2 = \lambda_1 S^2 + \lambda_2 K^2 + \lambda_3 P^2 \\ \text{s.t. } &\boldsymbol{D_r} \leq \boldsymbol{L} \leq \boldsymbol{D_l}, \end{aligned} \tag{3.83}$$

where $\boldsymbol{D_r}, \boldsymbol{D_l}$ represent the left border and right border, respectively, and $\lambda_1, \lambda_2, \lambda_3$ denote the weight for each term, respectively.

Lane Keeping

The lane-keeping maneuver is one of the most common driving tasks, assuming that the host vehicle is required to keep a desired lateral distance L_d to the centerline, then an extra quadratic term D^2 can be constructed as follows:

$$D^2 = \sum_{i=1}^{n}(L_i - L_d)^2 = \boldsymbol{L}^T\boldsymbol{L} - 2\boldsymbol{L}_d^T\boldsymbol{L} + \text{cost}. \tag{3.84}$$

Again, cost is a constant term that is independent on the control variable $\boldsymbol{L}$ and it can be omitted. Therefore, combining all indicators listed in Sections 3.4.1–3.4.3, as well as the lane keeping term, the optimal path planning for lane keeping is

$$\begin{aligned} &\min \; \Sigma^2 = \lambda_1 S^2 + \lambda_2 K^2 + \lambda_3 P^2 + \lambda_4 D^2 \\ &\text{s.t. } \boldsymbol{D_r} \leq \boldsymbol{L} \leq \boldsymbol{D_l}, \end{aligned} \tag{3.85}$$

where λ_4 denotes the weight for lane-keeping term.

Obstacle Avoidance

Extra constraints need to be added when dealing with the situation of obstacle avoidance. As shown in Figure 3.12, the nearest control point P_j to the obstacle is estimated by the distance gap D, plus the size of $[d/\Delta s]$ control points within a certain distance d before and after the nearest control point P_k, where $k = [D/\Delta s]$, Δs denotes the distance gap of adjacent control points, and $[\cdot]$ denotes the "floor" operation. The lateral distance to the centerline of those control points is at least a safety distance threshold r_{rs} greater than the lateral displacement of the obstacle, to ensure that the obstacle avoidance path does not collide with the obstacle.

Finally, the optimal obstacle avoidance path planning problem is described as a quadratic optimal planning problem shown below, that is,

$$\begin{aligned} &\min \; \Sigma^2 = \lambda_1 S^2 + \lambda_2 K^2 + \lambda_3 P^2 \\ &\text{s.t. } D_r \leq L_k \leq D_l, k = 1, 2, \dots, j - k - 1, j + k - 1, \dots, n + 1 \\ &\quad L_{obst} + r_s \leq L_k \leq D_l, k = j - k, \dots, j, \dots, j + k. \end{aligned} \tag{3.86}$$

Quadratic programming problem can be handled easily in MATLAB by using the function of *quadprog*. Finally, all path points in the Frenet coordinate space would transfer back to the Cartesian coordinate space via the relationship, as shown in Equation (3.4).

Case Studies

Quadratic programming problems can be handled easily in MATLAB by using the function of quadprog. Finally, all path points in the Frenet coordinate space would transfer back to the Cartesian coordinate space via the relationship, as shown in (3.4).

(1) *Free driving:* Figure 3.13 shows an example of free driving on a racing-like course, where the vehicle has to finish the course within the road boundaries. The approximate circumference of the course is around 4200 m, and the gap between adjacent nodes is 5 m, which is 840 nodes in total. We may find that the result of the planned path

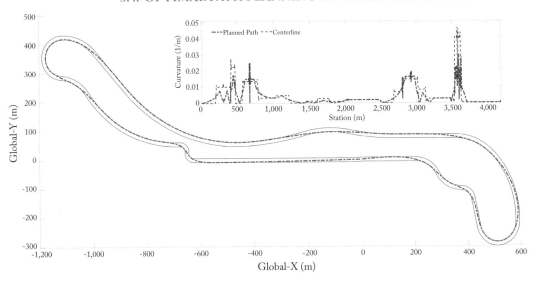

Figure 3.13: Path planning of free driving on a racing-like course.

is quite reasonable, especially followed by a maneuver of "outside-inside-outside" in continuous corners. Moreover, the planned path is much smoother compared with the centerline of the course.

(2) *Lane keeping:* Figure 3.13 shows an example of lane keeping on a course with three large turns, and the vehicle is expected to keep a lateral distance as 2 m to the road centerline while traveling along with its right lane. The approximate circumference of the course is around 1000 m, the gap between adjacent nodes is 5 m, which is 200 nodes in total. The planning result is quite satisfactory and the curvature of the planned path is smoother than the centerline.

(3) *Obstacle avoidance:* Figure 3.13 shows an example of path planning for obstacle avoidance. The approximate planning distance is 120 m, the gap between adjacent nodes is 5 m, which is 40 nodes in total. Also, assuming that the obstacle is in the same lane of the host vehicle at a distance of 50 m away and the lateral distance to the centerline 2 m (right). The longitudinal distance of the host vehicle at the starting position on the road is 20 m while the target point is set at a longitudinal distance of 140 m. Besides, the longitudinal safety distance around the host vehicle for obstacle avoidance d is 20 m and the safety distance threshold r_s is 2 m to ensure that the planned path avoiding the collision with the obstacle. As shown in Figure 3.13, the planned path with a natural shape can successfully avoid the obstacle ahead.

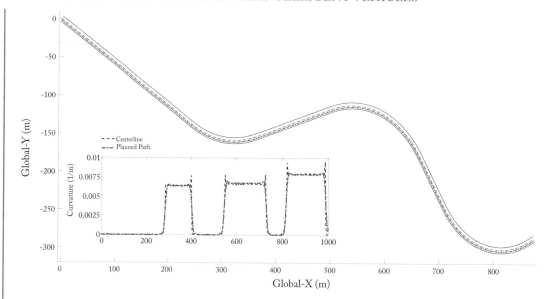

Figure 3.14: Path planning of lane keeping on a curved course with three turns.

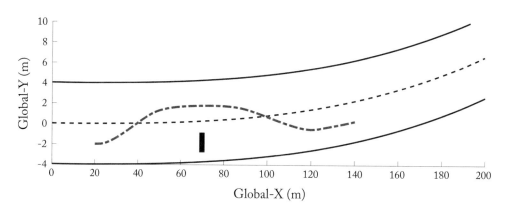

Figure 3.15: Path planning of obstacle avoidance on a curved course.

3.5   SPEED PLANNING UNDER THE NON-FOLLOWING SCENARIO

In this section, two kinds of speed-planning algorithms for non-following scenarios (e.g., free driving, lane change) are introduced, which includes the multi-object optimization based on high-order polynomials and the convex optimization based on the natural cubic splines.

3.5.1   SPEED PLANNING BY THE MULTI-OBJECT OPTIMIZATION BASED ON HIGH-ORDER POLYNOMIALS

Speed planning is quite important for the comfort and safety of the autonomous vehicles, especially when traveling along the route with large curvatures. Unlike most studies that use linear-varying speed profiles, a quartic polynomial will be used to describe the speed profile over time since the corresponding acceleration is continuous and be capable of representing conventional speed variations for some time. Then the longitudinal speed of the host vehicle $u(t)$ can be expressed as

$$u(t) = \sum_{k=1}^{5} k b_k t^{k-1}, \ t \in [0, t_f],\quad (3.87)$$

where $b_k \in \mathcal{B}(k = 1, \ldots, 5)$ denotes the polynomial coefficient, t_f denotes the end time, and the corresponding acceleration A can be further calculated by

$$A(t) = \frac{du(t)}{dt} = \sum_{k=1}^{5} k(k-1) b_k t^{k-2}, \ t \in [0, t_f].\quad (3.88)$$

Consider the initial speed u_0 and the initial acceleration a_{x0} measured by sensors, such that,

$$b_1 - u_0 = 0, b_2 - \frac{a_{x0}}{2} = 0.\quad (3.89)$$

Moreover, the jerk α is calculated by differentiating a_x respecting the t, which yields

$$\alpha(t) = \frac{da_x(t)}{dt} = \sum_{k=1}^{5} k(k-1)(k-2) b_k t^{k-3}, \ t \in [0, t_f].\quad (3.90)$$

Finally, the traveled distance s in the time interval $[0, t_f]$ is evaluated by

$$s(t) = \int_{\tau=0}^{\tau=t_f} u(\tau) d\tau = \sum_{k=1}^{5} b_k t^k, \ t \in [0, t_f].\quad (3.91)$$

Furthermore, there exist velocity and acceleration constraints for vehicle velocity planning, which are composed of the following terms. (1) Lateral acceleration bounds, which is simply calculated by

$$a_y = u^2 / \rho \leq a_{ymax}.\quad (3.92)$$

Then,

$$u \leq u_{max} = \sqrt{|a_{ymax}| \rho},\quad (3.93)$$

where u_{max} denotes the maximum allowed speed at the reference point, $a_{ymax} = \min\{\mu g, a_{yusr}\}$, μ denotes the road adhesion constant, and a_{yusr} denotes the maximum allowed lateral acceleration value expected by the decision-maker, and ρ denotes the road curvature radius. (2) To guarantee a pure rolling motion to avoid the phenomenon of wheel skidding, it is necessary to verify that forces transmitted to the ground are smaller than the ground friction force. For simplicity, the total tire force is evaluated by

$$F_t = \sqrt{F_x^2 + F_Y^2} = m\sqrt{a_x^2 + (u^2/\rho)^2} \leq \mu mg \tag{3.94}$$

such that

$$a_x^2 + u^4/\rho^2 \leq \mu^2 g^2. \tag{3.95}$$

Then,

$$|a_x| \leq \sqrt{\mu^2 g^2 - u^4/\rho^2}. \tag{3.96}$$

Therefore, the acceleration constraint is

$$|a_x| \leq \min\{\sqrt{\mu^2 g^2 - u^4/\rho^2}, a_{xusr}\}, \tag{3.97}$$

where a_{xusr} denotes the maximum allowed longitudinal acceleration value according to practical situations. Generally, the longitudinal acceleration of the host vehicle should be not too high to affect the driving comforts.

We choose a set of $N_p = [\frac{s_p}{\Delta s}]$ equivalently spaced reference points over a given preview path length s_p with an incremental Δs. The reaching time at each point is denoted by $t_j, j = 1, 2, \ldots, N$. Therefore, we could obtain a size of N_p equalities regarding the traveled station s_j,

$$\left\{ s_j(t_i) - \frac{j}{N_p} s_p = 0, j = 1, 2, \ldots, N \right\}. \tag{3.98}$$

Then the speed planar aims to maximize the velocity at each sampling point while minimizing the average speed change rate and jerk over the planning distance simultaneously to obtain good driving comforts as well, such that the objective function could be designed as

$$\min_{\{\mathcal{B}, \tau\} \in \mathcal{Z}} \begin{bmatrix} \frac{1}{u(t_i)} - \frac{1}{u_{\max}(t_i)}, i = 1, 2, \ldots, N_p \\ \frac{1}{N_p} \sum_{i=1}^{N_p} A(t_i)^2 \\ \frac{1}{N_p} \sum_{i=1}^{N_p} J(t_i)^2 \end{bmatrix}. \tag{3.99}$$

The control variable $\mathcal{Z}$ includes the coefficients of the velocity and the reached time of each point, namely

$$\mathcal{Z} = \begin{bmatrix} b_1 & b_2 & b_3 & b_4 & b_5 & t_1 & t_2 & \cdots & t_{N_p-1} & t_{N_p} \end{bmatrix}^T.$$

The problem as shown in Equation (3.99) is called the multi-objective programming in which the solution minimizes one function often while it does not minimize the others simultaneously. There is usually no unique optimal solution for them. However, the decision maker generally has a goal for each objective in mind. The objective functions are valued based on the priority, in that case, the goal programming method can be applied to solve this problem, which could be further converted to a Minimax problem defined in Equation (3.100) subject to the speed and acceleration bounds, namely,

$$
\min_{b_k, t \in \mathbf{Z}} \ \max
\begin{bmatrix}
w_i \left(\dfrac{1}{u(t_i)} - \dfrac{1}{u_{\max}(t_i)} - G_i \right) \\[2ex]
w_{N_p+1} \left(\dfrac{1}{N_p} \displaystyle\sum_{i=1}^{N_p} A(t_i)^2 - G_{N_p+1} \right) \\[2ex]
w_{N_p+2} \left(\dfrac{1}{N_p} \displaystyle\sum_{i=1}^{N_p} \alpha(t_i)^2 - G_{N_p+2} \right)
\end{bmatrix}
\tag{3.100}
$$

$$
\text{s.t.} \ \left\{ s_j(t_i) - \frac{j}{N_p} s_p = 0, \, b_1 - u_0 = 0, \, b_2 - \frac{a_{x0}}{2} = 0 \right\}
$$

$$
0 < u(t_i) \leq \sqrt{|a_{ymax}| \rho(t_i)}, \, a_{ymax} = \min\{\mu g, a_{yusr}\}
$$

$$
|a_x| \leq \min\{\sqrt{\mu^2 g^2 - u^4/\rho^2}, a_{xusr}\},
$$

where u_0, a_{x0} denote the initial speed and speed change rate of the host vehicle at the planning instant, respectively, G_i denotes the target value for each sub-objective function, and w_i denotes the weight for each sub-objective function.

Moreover, Equation (3.100) could be easily implemented and efficiently solved by the Matlab functions such as *fminimax* or *fgoalattain*, where the SQP algorithm is applied. The Sequential Quadratic Programming (SQP) method represents the state-of-the-art in nonlinear programming domains, which are particularly effective when the objective and constraint functions are expensive to evaluate. The SQP method solves the Nonlinear Programming (NLP) problem via a sequence of QP subproblems. The constraints of each QP subproblem are the linearization of the nonlinear constraints in the original problem, and the subproblem objective function is a quadratic approximation to a modified Lagrangian function. The solution of each subproblem defines a search direction, and the active set of constraints (at the solution) provides an estimate of the active set for the original problem.

3.5.2 SPEED PLANNING BY THE NATURAL CUBIC SPLINE AND CONVEX OPTIMIZATION

When the host vehicle moves in a non-following mode, then the objective of the speed planning is to generate speed profiles to achieve an optimal time consuming, as well as the driving comforts within the safety constraints. Similar to the optimal path planning in the previous section, we also assume that the velocity of the adjacent nodes was connected via natural cubic splines to obtain smooth and feasible speed profiles, namely,

$$u(s^*) = c_{0i} + c_{1i}s^* + c_{2i}s^{*2} + c_{3i}s^{*3}. \tag{3.101}$$

We consider the following criteria to ensure the vehicle moves in a safe and comfortable status.

Maximum Allowed Velocity

It keeps the host vehicle moving within the maximum allowed speed to save the traveling time, which could be denoted by a squared term as

$$V^2 = \sum_{i=1}^{N_p} (u(t_i) - u_{\max}(t_i))^2 = \boldsymbol{u}^T \boldsymbol{u} - 2\boldsymbol{u}_{\max}^T \boldsymbol{v} + \text{cost}, \tag{3.102}$$

where $\boldsymbol{u} = [u_0 \ u_1 \ \dots \ U_{n-1} \ u_n]^T$, and $u_{\max}$ is calculated based on the road curvature, namely,

$$\boldsymbol{u}_{\max} = \sqrt{(\text{diag}[\boldsymbol{K}])^{-1}\mu g}, \tag{3.103}$$

where $\boldsymbol{K}$ denotes the path curvature, g denotes the gravitational constant, and μ denotes the road adhesion coefficient.

Minimum Acceleration

It ensures the derivative of the velocity is small enough to achieve a good driving comfort, remembering that the first derivative for a natural cubic spline computed at $s^* = 0$ could be denoted by a linear form with a constant matrix $\mathbf{G}$, namely $\left.\dfrac{du(s^*)}{ds^*}\right|_{s^*=0} = \mathbf{G}\boldsymbol{u}$; finally, it is denoted by a squared term as

$$A^2 = \sum_{i=1}^{n}(\Delta s A_i)^2 = \Delta s^2 \sum_{i=1}^{n} \left(\frac{d^2 u_{xi}(s^*)}{ds^*}\right)^2 = \boldsymbol{u}^T \mathbf{G}^T \mathbf{G}\boldsymbol{u}. \tag{3.104}$$

Minimum Jerk

It ensures the derivative of the longitudinal acceleration is small enough to achieve a good driving comfort, and the second derivative for a natural cubic spline computed at $s^* = 0$ could be denoted by a linear form with a constant matrix G, namely $\left.\dfrac{du(s^*)}{ds^*}\right|_{s^*=0} = \mathbf{H}\boldsymbol{u}$. Such that the jerk term can be denoted as a squared term,

$$J^2 = \sum_{i=1}^{n}(\Delta s J_i)^2 = \Delta s^4 \sum_{i=1}^{n} \left(\frac{d^2 u_{xi}(s^*)}{ds^*}\right)^2 = \boldsymbol{u}^T \mathbf{H}^T \mathbf{H}\boldsymbol{u}. \tag{3.105}$$

Finally, the combined cost function for velocity planning is defined by

$$\Sigma^2 = \varepsilon_1 V^2 + \varepsilon_2 A^2 + \varepsilon_3 J^2 = \boldsymbol{u}^T (\varepsilon_1 \mathbf{I} + \varepsilon_2 \mathbf{G}^T \mathbf{G} + \varepsilon_3 \mathbf{H}^T \mathbf{H}) \boldsymbol{u} - 2\varepsilon_1 \boldsymbol{v}_{\max}^T \boldsymbol{u}, \qquad (3.106)$$

where $\varepsilon_1, \varepsilon_2, \varepsilon_3$ denotes weight constants for each quadratic term.

As for the constraints of the ground, which is read by $a_x^2 + K^2 u^4 \leq \mu^2 g^2$, further, it could be expanded and expressed as a matrix form for all reference points, which is

$$(\mathbf{G}\boldsymbol{u}_x)^2 - \mu^2 \boldsymbol{g}^2 + \text{diag}([K_i^2])[u_i^4] \leq 0. \qquad (3.107)$$

Finally, the velocity planning could be interpreted by a convex optimization problem as well, namely,

$$\min \; \Sigma^2 = \boldsymbol{u}^T (\varepsilon_1 \mathbf{I} + \varepsilon_2 \mathbf{G}^T \mathbf{G} + \varepsilon_3 \mathbf{H}^T \mathbf{H}) \boldsymbol{u} - 2\varepsilon_1 \boldsymbol{v}_{\max}^T \boldsymbol{u} \qquad (3.108\text{a})$$

$$\text{s.t.} \; \Delta \boldsymbol{u}_{\min} \leq \Delta \boldsymbol{u} = \mathbf{W}\boldsymbol{u} - u_0 E \leq \Delta \boldsymbol{u}_{\max} \qquad (3.108\text{b})$$

$$(\mathbf{G}\boldsymbol{u}_x)^2 - \mu^2 \boldsymbol{g}^2 + \text{diag}([K_i^2])[u_i^4] \leq 0 \qquad (3.108\text{c})$$

$$\boldsymbol{u}_{\min} \leq \boldsymbol{v}_x \leq \boldsymbol{u}_{\max}, \qquad (3.108\text{d})$$

where

$$\mathbf{W} = \begin{bmatrix} 1 & 0 & \cdots & 0 & 0 \\ -1 & 1 & \cdots & 0 & 0 \\ \vdots & \vdots & \ddots & \vdots & \vdots \\ 0 & 0 & \cdots & -1 & 1 \end{bmatrix}_{(n+1)\times(n+1)} \qquad E = \begin{bmatrix} 1 \\ 0 \\ \vdots \\ 0 \end{bmatrix}_{(n+1)\times 1}.$$

The solution search could be realized by *fmincon* function in Matlab, and the active-set method could be applied for solving the convex optimization problem.

3.5.3   CASE STUDY

Figure 3.16 lists two examples of the velocity planning for different scenarios by using the multi-objective optimization technical; it includes global positions of the road centerline where the first course is a curved one with three large turns, while the second course with a shape of an "S" is composed of two large arc turns. The user-defined maximum lateral acceleration and longitudinal acceleration are ± 0.4 g, ± 0.2 g, respectively. The vehicle has an initial velocity of 90 km/h, the speed-planning distance is 50 m and the number of reference points at each planning period is 10. As shown in Figure 3.16, besides the planned speed, the planning results also include the information of the corresponding accelerations and jerks. Overall, the planning results are satisfying; we can observe that the planned velocity lie within the speed constraints, the velocity profile is feasible, the accelerations are smooth at the vast majority of places, and the jerks are not large, which are acceptable. Figure 3.17 shows another example of speed planning for the planned path in the racing-like course, as shown in Figure 3.13, where the racing car has to finish one lap as fast as it can, however, try to avoid large accelerations/decelerations. For

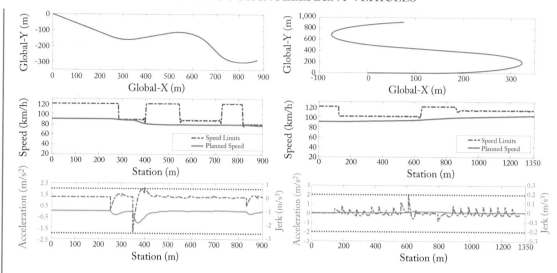

Figure 3.16: Speed-planning demos on different road shapes by the multi-object optimization with high-order polynomials.

safety purposes, let us assume the maximum speed of the racing car is 40 m/s, and the desired lateral acceleration should be less than 1.0 g, moreover, the maximum increment of the speed rate respecting to the station is 0.2 s^{-1} and the minimum increment of the speed rate respecting to the station is -0.4 s^{-1}. The planning distance gap between adjacent modes is 5 m which is 850 nodes in total, and the weight for the speed-planning objective function $\boldsymbol{w} = [0.8, 0.1, 0.1]$. Finally, the speed planned by the natural cubic spline and convex optimization, as shown in Figure 3.17, is quite good, the speed profiles are smooth, and the planned speed and the speed rate also lie within the bounds.

3.6 SPEED PLANNING IN THE CASE OF CAR FOLLOWING

The velocity planning method introduced in Section 3.5 is only suitable when the host vehicle's velocity is independent of surrounding vehicles on the road. It will not work when the host vehicle follows a leading vehicle in front since the speed would change accordingly with the leading vehicle. Thus, we shall develop a velocity-planning algorithm for the following scenario. Technically, it would be very similar to the Adaptive Cruise Control. Assuming that the host vehicle detects a leading vehicle ahead, according to the kinematic relationship between the host vehicle and the leading vehicle as illustrated in Figure 3.18, the headway error Δd and relative speed error Δv could be determined by

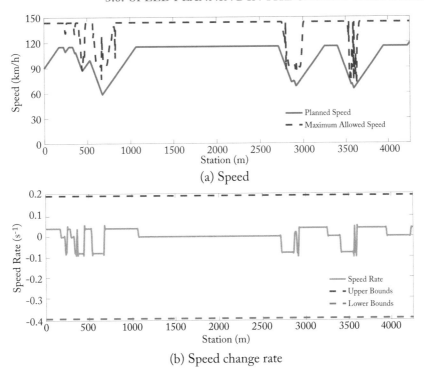

(a) Speed

(b) Speed change rate

Figure 3.17: Speed planning for the race map by the convex optimization with natural cubic splines.

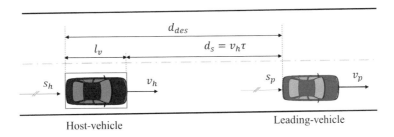

Figure 3.18: Illustration of adaptive following.

$$\begin{cases} \Delta d = d_{des} - (s_p - s_h) = (l_v + v_h \tau_d) - (s_p - s_h) \\ \Delta v = v_p - v_v, \end{cases} \qquad (3.109)$$

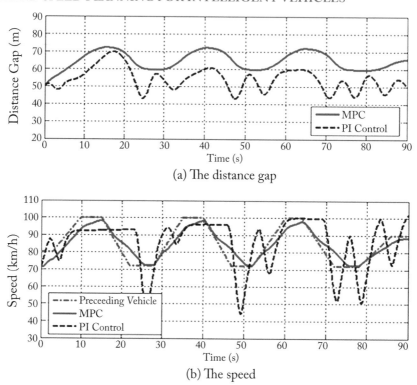

(a) The distance gap

(b) The speed

Figure 3.19: Speed-planning demos on the car-following scenario.

where the length of the vehicle is l_v, v_p, v_h denote the longitudinal speed of the leading vehicle and the host vehicle, respectively, τ_d is the desired time headway to the leading vehicle, while s_p, s_h denote the traveled distance of the leading vehicle and the host vehicle, respectively.

Relationship of Equation (3.109) could lead to a dynamical system as

$$\begin{bmatrix} \dot{\Delta d} \\ \dot{\Delta v} \end{bmatrix} = \begin{bmatrix} 0 & -1 \\ 0 & 0 \end{bmatrix} \begin{bmatrix} \Delta d \\ \Delta v \end{bmatrix} + \begin{bmatrix} \tau \\ -1 \end{bmatrix} a_h + \begin{bmatrix} 0 \\ 1 \end{bmatrix} a_p, \tag{3.110}$$

where a_p, a_h denote the longitudinal acceleration of the leading vehicle and the host vehicle, respectively. Finally, Equation (3.109) can be expressed by a state-space form $\dot{x} = f(x, u)$, where $x = [\Delta d \ \Delta v]^T$ is the state of the system, and $u = a_h$ represents the control input, while the acceleration of the leading vehicle a_p would be regarded as a disturbance to the system. Irrespective of the adaptive following control, the host vehicle is supposed to maintain a safe headway to the leading vehicle with minimum energy costs. Thus, a cost function related to the headway error, the relative speed error, and the longitudinal acceleration of the host vehicle is defined as

$$J_f = \sum_{i=1}^{n_p}(w_1\Delta d^2(k+i|k) + w_2\Delta v^2(k+i|k)) + \sum_{i=0}^{n_c-1} w_3 a_h^2(k+i|k), \qquad (3.111)$$

where w_1, w_2, w_3 are the weight constants for headway error, relative speed error, and the longitudinal acceleration of the host vehicle, respectively; and n_p, n_c denote the prediction horizon and control horizon in the MPC algorithm, respectively. Finally, the solution finding for minimizing Equation (3.111) could be solved as a convex optimization problem, namely,

$$\begin{aligned} \min_{a_h} \ & J_f \\ \text{s.t. } \boldsymbol{x}(k+i+1|k) = & \boldsymbol{f}(k+i|k, k+i|a_h), i = 0, \dots, n_p - 1 \\ a_{\min} \leq & a_h \leq a_{\max} \\ \Delta a_{\min} \leq & \Delta a_h \leq \Delta a_{\max}. \end{aligned} \qquad (3.112)$$

A simple example is shown below to show the feasibility and the effectiveness of the MPC for speed planning in the car-following scenario. The scenario describes that the ego vehicle follows the leading vehicle in front on a straight road, the desired time headway to the leading vehicle is set to 2.5 s. The initial speed of the ego vehicle is 72 km/h and the initial distance gap is 50 m, while the initial speed of the leading vehicle is 80 km/h, however, as shown in Figure 3.19b, its speed will accelerate to 100 km/h gradually, then drop it down to 72 km/h after some time, repeat this procedure with three cycles, and finally, accelerate to 90 km/h. The predictive and control horizon for the MPC is set to 10, and the maximum and minimum acceleration for the ego vehicle is 1.5 m/s^2 and -3 m/s^2, respectively.

The simulation results, as shown in Figure 3.19, include the speed of the ego vehicle and the distance gap between the ego vehicle and the leading vehicle. We compare it with the planning results of the well-known PI method to highlight the advantages of the MPC. Based on the simulation results, it can be observed that the MPC can adapt to the speed change of the leading vehicle more sensitively compared with the PI method; besides, the distance gap can be maintained well to match the optimal one. Moreover, the speed profiles planned by the MPC would not be dramatically changed and the driving comfort of the ego vehicle is satisfying.

3.7   CONCLUSIONS

In this chapter, we mainly introduce path planning algorithms of elastic bands and its improvements based on the artificial potentials for obstacle avoidance, which is feasible for any shapes of the structured road due to the help of road Frenet coordinates. Besides, the path-planning method based on the combination of the convex optimization and the natural cubic splines for different driving scenarios, such as lane keeping, free driving, and obstacle avoidance, is also proposed and validated in this chapter. Speed-planning methods for the non-following scenario are elaborated in the subsequent contents, which includes the multi-object optimization-based speed-planning method with high-order polynomials, and similarly, the combination of the convex optimization and the natural cubic splines for obtaining optimal speed profiles with satisfying driving comforts and sufficient safety. While for the case of car following, the MPC method based on the car-following kinematic mechanisms is applied to adaptively following the leading vehicle.

CHAPTER 4

Robust Trajectory Tracking Methods for Intelligent Vehicles

In this chapter, at first, we will separately introduce robust tracking control methods on speed or path profiles based on the adaptive SMC, then followed by an integrated controller to track them both via the linearized MPC, which is based on a 5DOF nonlinear vehicle model. Related study cases will be elaborated to show the feasibility and effectiveness of those methods.

4.1 LONGITUDINAL VELOCITY-TRACKING CONTROLLER

We will present a robust SMC controller based on the input-output linearization method to track the planned velocity precisely, validations, and comparisons with some classical methods that are also conducted to show the advantages. At first, let us consider the longitudinal motion for the driven wheel of the vehicle, as shown in Figure 4.1, which is governed by,

$$m\dot{u} = F_x - F_{roll} - F_{wind}, \tag{4.1}$$

where F_x is the tire longitudinal force, and it is a function of the vertical tire load, the longitudinal tire slip, and the friction coefficient.

Moreover, the longitudinal tire force could be expressed as a linear form in normal driving situations where the acceleration or deceleration of the vehicle is not large, namely

$$F_x = C_x\kappa = C_x\frac{r\omega_w - u}{\max(r\omega_w, u)} = \begin{cases} C_x(\frac{r\omega_w}{u} - 1), & \text{if } r\omega_w \leq u \\ C_x(1 - \frac{r\omega_w}{u}), & \text{if } r\omega_w > u, \end{cases} \tag{4.2}$$

where κ denotes the longitudinal slip, ω_w denotes the angular speed of the tire, and C_x denotes the longitudinal stiffness of the tire. Generally, the rolling resistance which is much smaller than the longitudinal force is then neglected ($F_{roll} \ll F_x$), and the wind resistance is proportional with the square of the velocity, therefore, Equation (4.2) could be rewritten as

$$\dot{u} = \frac{C_x}{m}\frac{r\omega_w - u}{\max(r\omega_w, u)} - \frac{\rho C_{rx} A_r u^2}{2m}. \tag{4.3}$$

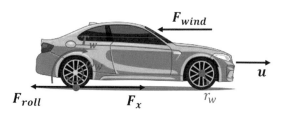

Figure 4.1: The longitudinal motion of the vehicle.

The wheel dynamic equation is then written by the following:

$$\dot{\omega}_w = \frac{T_w - F_x r}{J_w} = \frac{T_w}{J_w} - \frac{C_x r}{J_w} \frac{r\omega_w - u}{\max(r\omega_w, u)}, \tag{4.4}$$

where T_w represents the control torque acting on the driving wheel. Combining Equations (4.3) and (4.4), the longitudinal motion is written as a state-space form as

$$\begin{cases} \begin{bmatrix} \dot{u} \\ \dot{\omega}_w \end{bmatrix} = \begin{bmatrix} \frac{C_x}{m} \frac{r\omega_w - u}{\max(r\omega_w, u)} - \frac{\rho C_{rx} A_r u^2}{2m} \\ -\frac{C_x r}{J_w} \frac{r\omega_w - u}{\max(r\omega_w, u)} \end{bmatrix} + \begin{bmatrix} 0 \\ \frac{1}{J_w} \end{bmatrix} T_w. \\ p = u \end{cases} \tag{4.5}$$

Let us take care of the system (4.5) for braking condition that $r\omega_w < u$ first, and set,

$$\frac{C_x r}{m} = c_{10}, \frac{C_x}{m} = c_{11}, \frac{\rho C_{rx} A_r}{2m} = c_{12}, -\frac{C_x r^2}{J_w} = c_{20}, -\frac{C_x r}{J_w} = c_{21}, \frac{1}{J_w} = c_{22}. \tag{4.6}$$

Therefore, rewrite the dynamic system (4.5) as the form

$$\begin{cases} \dot{x} = f(x) + g T_w \\ p = h(x) \end{cases} \tag{4.7}$$

with

$$x = \begin{bmatrix} u \\ \omega_w \end{bmatrix}, f = \begin{bmatrix} c_{10} \frac{\omega_w}{u} - c_{11} - c_{12} u^2 \\ c_{20} \frac{\omega_w}{u} - c_{21} \end{bmatrix}, g = \begin{bmatrix} 0 \\ c_{22} \end{bmatrix}, h(x) = u.$$

Noting that the system Equation (4.7) is a nonlinear one, and the common control method for the linear system does not work anymore, however, it is easy to verify that the system processes a relative order of 2, thus the input-output linearization approach could be applied. There exists a diffeomorphism, $T : D \subset \mathbb{R}^2 \to \mathbb{R}^2$, and a coordinate transformation as

$$\begin{bmatrix} z_1 \\ z_2 \end{bmatrix} = T(x) = \begin{bmatrix} h(x) \\ L_f h(x) \end{bmatrix} = \begin{bmatrix} u \\ c_{10} \frac{\omega_w}{u} - c_{11} - c_{12} u^2 \end{bmatrix}. \tag{4.8}$$

Furthermore, Equation (4.8) could realize the system's input-output linearization, since

$$\begin{cases} h(x) = u \\ L_g L_f h(x) = \dfrac{c_{10} c_{22}}{u} \\ L_f^2 h(x) = \dfrac{2c_{12}^2 u^6 + 2c_{11}c_{12}u^4 - c_{12}c_{10}\omega_w u^3 - c_{21}c_{10}u^2 + c_{10}(c_{11} + c_{20}\omega_w u) - c_{10}^2 \omega_w^2}{u^3}. \end{cases}$$

The system could be reformed as

$$\begin{bmatrix} \dot{z}_1 \\ \dot{z}_2 \end{bmatrix} = \begin{bmatrix} 0 & 1 \\ 0 & 0 \end{bmatrix} \begin{bmatrix} z_1 \\ z_2 \end{bmatrix} + \begin{bmatrix} 0 \\ 1 \end{bmatrix} L_g L_f h(x) \left(T_w + \dfrac{L_f^2 h(x)}{L_g L_f h(x)} \right). \tag{4.9}$$

Let $\Theta(x) := L_f^2 h(x)$, $\Lambda(x) := L_g L_f h(x)$. Noting that we have the output $u = z_1$, thus,

$$\ddot{u} = \Theta(x) + \Lambda(x) T_w. \tag{4.10}$$

Moreover, we add an extra term in Equation (4.10) to make the system more robust to uncertainties caused by model inaccuracy and external disturbances, such that

$$\ddot{u} = \Theta(x) + \Lambda(x) T_w + \Delta, \tag{4.11}$$

where Δ denotes the total uncertainties, assuming that the longitudinal reference speed be a continuously differentiable signal u_{ref}, then the velocity tracking error is defined by

$$e_u = u - u_{ref}. \tag{4.12}$$

The sliding surface S of the first-order SMC is denoted as

$$S = \dot{e}_u + c e_u. \tag{4.13}$$

Still, let us definite a positive Lyapunov function as $V_0 = \frac{1}{2}S^2$; The stability condition and asymptotic convergence to the surface $S = 0$ are achieved by the following η-attractive condition, which constrains the trajectory of the system to side along the hypersurface S, namely,

$$\dot{V}_0 = S\dot{S} = S(\ddot{e}_u + c\dot{e}_u) \le S(\Theta(x) + \Lambda(x)T_w + \Delta - \ddot{u}_{ref} + c\dot{e}_u). \tag{4.14}$$

Let

$$\Theta(x) + \Lambda(x)T_w + \Delta - \ddot{u}_{ref} + c\dot{e}_u = -\eta \operatorname{sign}(S). \tag{4.15}$$

Then the control law is then chosen as

$$T_w = \dfrac{\ddot{u}_{ref} - c\dot{e}_u - \Theta(x) - \Delta_m - \eta \operatorname{sign}(S)}{\Lambda(x)}, \tag{4.16}$$

where Δ_m denotes the maximum value of the total uncertainty, such that

$$\dot{V}_0 = S\dot{S} = -\eta \operatorname{sign}(S) + (\Delta - \Delta_m) = -\eta |S| + (\Delta - \Delta_m) < 0, \text{ for } \eta > 0. \tag{4.17}$$

Although $|\Delta| < |\Delta_m|$, it is not easy to determine the upper bound of $\Delta_{\min}$ real applications, it is necessary to develop an adaptive controller for it. Assuming that Δ_m is changing slowly, which means,

$$\dot{\Delta}_m = 0. \tag{4.18}$$

Moreover, define a second Lyapunov function as,

$$V_1 = V_0 + \frac{1}{2\varepsilon}\Delta\epsilon^2, \tag{4.19}$$

where $\varepsilon > 0$, $\Delta\epsilon = \hat{\Delta}_m - \Delta_m$ is the error between Δ_m and its estimation $\hat{\Delta}_m$. Since $\dot{\Delta}_m = 0$, such that,

$$\dot{V}_1 = \dot{V}_0 + \frac{1}{\varepsilon}\Delta\epsilon\dot{\Delta}\epsilon \leq S(\Theta(x) + \Lambda(x)T_w + \Delta_m - \ddot{u}_{ref} + c\dot{e}_u) + \frac{1}{\varepsilon}\Delta\epsilon\dot{\Delta}\epsilon. \tag{4.20}$$

And set the control law as,

$$T_w = \frac{\ddot{u}_{ref} - c\dot{e}_u) - \Theta(x) + \Lambda(x)T_w + \hat{\Delta}_m - \eta\,\text{sign}(S)}{\Lambda(x)}. \tag{4.21}$$

Substitute Equation (4.21) into (4.20), then,

$$\dot{V}_1 \leq \Delta\epsilon\left(-S + \frac{1}{\varepsilon}\dot{\Delta}\epsilon\right) - \eta\,\text{sign}(S). \tag{4.22}$$

Therefore, if we set the adaptive control law as

$$\dot{\Delta}\epsilon = \varepsilon S(\varepsilon > 0). \tag{4.23}$$

Again, to eliminate the chattering effect, the discontinuous component $\text{sign}(S)$ is replaced by a continuous function $\Pi(S)$ defined by,

$$\Pi : S \to \frac{S}{|S| + \sigma}.$$

The completed control law (4.21) rewrites as follows,

$$T_w = \frac{\ddot{u}_{ref} - c\dot{e}_u) - \Theta(x) + \Lambda(x)T_w + \hat{\Delta}_m - \eta\frac{S}{|S| + \sigma}}{\Lambda(x)} \tag{4.24}$$

with

$$\begin{cases} \Theta(x) = \dfrac{2c_{12}^2u^6 + 2c_{11}c_{12}u^4 - c_{12}c_{10}\omega_w u^3 - c_{21}c_{10}u^2 + c_{10}(c_{11} + c_{20}\omega_w u) - c_{10}^2\omega_w^2}{u^3} \\ \Lambda(x) = \dfrac{c_{10}c_{22}}{u}, \end{cases}$$

where $\hat{\Delta}_m$ is subjecting to the dynamical system (4.23). Finally, substituting Equation (4.24) into (4.20), we have

$$\dot{V}_1 \leq S\left(-\Delta\epsilon - \eta\frac{S}{|S|+\sigma}\right)\frac{1}{\varepsilon}\Delta\epsilon\,\dot{\Delta}\epsilon = \Delta\epsilon\left(-S + \frac{1}{\varepsilon}\Delta\epsilon\,\dot{\Delta}\epsilon\right) - \frac{\eta S^2}{|S|+\sigma} = -\frac{\eta S^2}{|S|+\sigma} < 0$$

(4.25)

which is held for any $\eta > 0, \sigma > 0$, meaning that control law (4.24) could stabilize the tracking system. Similarly, we can derive a control law in an accelerating situation, besides, noting that the torque acting on the wheel might be a driving torque for acceleration ($T_w > 0$) or braking torque for decelerating ($T_w < 0$), and if T_w is the driving torque, then the engine torque T_e and its angular speed ω_e could be approximated by

$$\begin{cases} T_w = T_w/(R_{gear}R_{diff}\eta) \\ \omega_e = \omega_w R_{gear}R_{diff}, \end{cases}$$

(4.26)

where R_{gear} denotes the ratio of the gearbox, η denotes the efficacy of the gearbox, and R_{diff} denotes the ratio of the differential. Therefore, the throttle γ of the engine system can be simply evaluated by looking up a 2D table of the engine throttle position respect to the engine speed and the engine output torque T_e, namely,

$$\gamma = \text{MAP}_\gamma(T_e, \omega_e) \text{ for } T_w \geq 0.$$

(4.27)

To verify the effectiveness and feasibility of the SMC-based longitudinal velocity tracking controller, it will be compared by a cubic PI as shown in (4.28), as well as a classical PI controller (when $k_{p3} = 0$).

$$a_x = k_{p3}(u - u_d)^3 + k_p(u - u_d) + k_i\int(u - u_d)dt,$$

(4.28)

where k_{p3}, k_p, and k_i are the design parameters of the cubic-PI controller.

The testing scenario is a vehicle with rather fierce velocity changes on a straight road. The vehicle accelerates from 60 km/h at the beginning, then with a gradual increment of 10 km/h, finally accelerates to 100 km/h, then it decelerates gradually to 60 km/h again, as shown the reference velocity in Figure 4.2a. The tracking results in Figure 4.2a show that the ego vehicle can follow the reference velocity well under the robust SMC control, the cubic PI, and regular PI, and the tracking results of the SMC is the best among them. Besides, we should also be aware that the cubic PI controller or regular PI is quite sensitive to its parameter selecting, which should be chosen carefully based on lots of simulations to achieve a good tracking performance, while the SMC controller would not suffer it. Additionally, we also list the normalized throttle control input of each controller in Figure 4.2b, which clearly shows that the control process is acceptable. Therefore, the SMC controller for vehicle velocity tracking is effective and feasible.

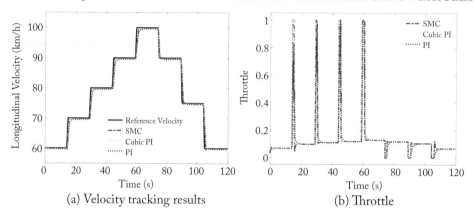

(a) Velocity tracking results

(b) Throttle

Figure 4.2: Longitudinal velocity tracking results with different control methods.

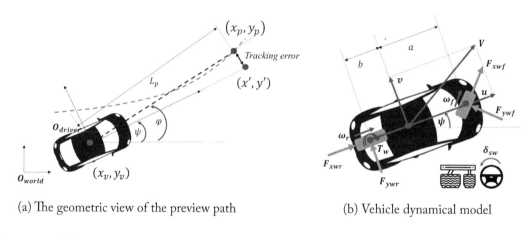

(a) The geometric view of the preview path

(b) Vehicle dynamical model

Figure 4.3: The geometric view of the preview path and vehicle model.

4.2   ROBUST SMC STEERING CONTROLLER

We introduce a robust adaptive preview-time SMC steering controller for path tracking in this section. For the preview point (x_p, y_p), according to the geometric relationship with the current vehicle position (x_v, y_v) as depicted in Figure 4.3, we have

$$\begin{cases} x_p = x_v + L_p \cos \varphi \\ y_p = y_v + L_p \sin \varphi, \end{cases} \tag{4.29}$$

where $L_p = uT_p$ denotes the preview distance, T_p denotes the preview time for the driver model, and φ denotes the heading of the vehicle, which could be approximated by the vehicle yaw angle ϕ when its sideslip is small. Thus, a dynamical system respecting to the preview position could be evaluated by

$$\begin{cases} \dot{x}_p = u\cos\phi - v\sin\phi - L_p r\sin\phi \\ \ddot{x}_p = \dot{u}\cos\phi - ur\sin\phi - \dot{v}\sin\phi - vr\cos\phi - L_p\sin\phi\dot{r} - L_p r^2\cos\phi \\ \dot{y}_p = v\cos\phi + u\sin\phi + L_p r\cos\phi \\ \ddot{y}_p = \dot{v}\cos\phi - vr\sin\phi + \dot{u}\sin\phi + ur\cos\phi + L_p\cos\phi\dot{r} - L_p r^2\sin\phi, \end{cases} \quad (4.30)$$

where u, v, r denotes the longitudinal speed, lateral speed, and yaw rate of the vehicle, respectively, and $\dot{u}$ denotes the longitudinal acceleration of the vehicle. Those are available from the on-board sensors of the vehicle (e.g., the IMU).

Set,

$$v_1 = \dot{u}\cos\phi - ur\sin\phi - vr\cos\phi - L_p r^2\cos\phi$$
$$v_2 = \dot{u}\sin\phi - vr\sin\phi + ur\cos\phi - L_p r^2\sin\phi.$$

Then, (4.30) could be rewritten as

$$\begin{cases} \ddot{x}_p = -\dot{v}\sin\phi - L_p\sin\phi\dot{r} + v_1 \\ \ddot{y}_p = \dot{v}\cos\phi + L_p\cos\phi\dot{r} + v_2. \end{cases} \quad (4.31)$$

Recalling a 2DOF vehicle model considering the model uncertainties could be represented by

$$\begin{cases} \dot{v} = (a_{11} + \Delta_{11})v + (a_{12} + \Delta_{12})r + (b_1 + \Delta_{13})\delta_{sw} \\ \dot{r} = (a_{21} + \Delta_{21})v + (a_{22} + \Delta_{22})r + (b_2 + \Delta_{23})\delta_{sw} \end{cases} \quad (4.32)$$

with

$$a_{11} = \frac{-2(C_f + C_r)}{mu}, \ a_{12} = \frac{-2aC_f - bC_r}{mu} - u, \ b_1 = \frac{C_f}{mG}$$

$$a_{21} = \frac{-2(aC_f - bC_r)}{I_z u}, \ a_{22} = \frac{-2(a^2 C_f + b^2 C_r)}{I_z u}, \ b_2 = \frac{2aC_f}{I_z G},$$

where C_f, C_r denotes the front/rear tire lateral stiffness, a, b denotes the front/rear distance from the Center of Gravity (CoG) to front/rear axle, m denotes the mass of the vehicle, I_z denotes the moment of inertia respecting the Z-axis of the vehicle, G denotes the steering ratio of the vehicle, and Δ represents the model uncertainties that could be caused by vehicle model nonlinearity or unmatched vehicle system parameters.

Rewrite (4.32) as

$$\begin{cases} \dot{v} = a_{11}v + a_{12}r + b_1\Delta_{13}\delta_{sw} + F_1 \\ \dot{r} = a_{21}v + a_{22}r + b_2\delta_{sw} + F_2, \end{cases} \quad (4.33)$$

where $F_i = \Delta_{i1}v + \Delta_{i2} + \Delta_{i3}\delta_{sw}$ represents the total uncertainties. As depicted in Figure 4.3, the SMC controller is expected to ensure the lateral error between the estimating point and the reference point be minimized. The lateral preview error is defined as

$$e_{yp} = (y_p - y_{refp})\cos\phi - (x_p - x_{refp})\sin\phi \quad (4.34)$$

such that

$$
\dot{e}_{yp} = \left((\dot{y}_p - \dot{y}_{refp}\cos\phi - (y_p - y_{refp}\sin\phi r))\right)
$$
$$
- \left((\dot{x}_p - \dot{x}_{refp}\sin\phi + (x_p - x_{refp}\cos\phi r))\right) \tag{4.35}
$$
$$
\ddot{e}_{yp} = (\ddot{y}_p\cos\phi - (y_p - y_{refp})\sin\phi\dot{r} - \ddot{y}_{refp}\cos\phi - 2(\dot{y}_p - \dot{y}_{refp})\sin\phi r
$$
$$
- (y_p - y_{refp})\cos\phi r^2) - (\ddot{x}_p\sin\phi + (x_p - x_{refp})\cos\phi\dot{r} - \ddot{x}_{refp}\sin\phi
$$
$$
+ 2(\dot{x}_p - \dot{x}_{refp})\cos\phi r - (x_p - x_{refp})\sin\phi r^2). \tag{4.36}
$$

Let

$$
\begin{aligned}
m_1 &= -\ddot{y}_{refp}\cos\phi - 2(\dot{y}_p - \dot{y}_{refp})\sin\phi r - (y_p - y_{refp})\cos\phi r^2 \\
m_2 &= -\ddot{x}_{refp}\sin\phi + 2(\dot{x}_p - \dot{x}_{refp})\cos\phi r - (x_p - x_{refp})\sin\phi r^2 \\
n_1 &= (y_p - y_{refp})\sin\phi \\
n_2 &= (x_p - x_{refp})\cos\phi.
\end{aligned} \tag{4.37}
$$

Substitute (4.37) in (4.36), yielding

$$
\begin{aligned}
\ddot{e}_{yp} &= \ddot{y}_p\cos\phi - n_1\dot{r} + m_1 - (\ddot{x}_p\sin\phi + n_2\dot{r} + m_2) \\
&= (a_{11}v + a_{12}r + (L_p - n_1 - n_2)(a_{21}v + a_{22}r)) + (b_1 + (L_p - n_1 - n_2))b_2)\delta_{sw} \\
&\quad + (F_1 + (L_p - n_1 - n_2)F_2) + v_2\cos\phi - v_1\sin\phi + m_1 - m_2.
\end{aligned} \tag{4.38}
$$

Further, to simplify the expression (4.38) let

$$
\begin{aligned}
\mathcal{R} &= v_2\cos\phi - v_1\sin\phi + m_1 - m_2 \\
\mathcal{M} &= a_{11}v + a_{12}r + (L_p - n_1 - n_2)(a_{21}v + a_{22}r) \\
\mathcal{N} &= b_1 + (L_p - n_1 - n_2)b_2 \\
\mathcal{T} &= F_1 + (L_p - n_1 - n_2)F_2.
\end{aligned} \tag{4.39}
$$

Finally, (4.38) becomes

$$
\ddot{e}_{yp} = \mathcal{M} + \mathcal{N}\delta_{sw} + \mathcal{R} + \mathcal{T}. \tag{4.40}
$$

Thus, the sliding surface S of the SMC is designed as

$$
S = \dot{e}_{yp} + ke_{yp}. \tag{4.41}
$$

Define a Lyapunov function as

$$
V_1 = \frac{1}{2}S^2. \tag{4.42}
$$

Thus,

$$
\dot{V}_1 = S\dot{S} = S(\ddot{e}_{yp} + k\dot{e}_{yp}) = S(\mathcal{M} + \mathcal{N}\delta_{sw} + \mathcal{R} + \mathcal{T}), \tag{4.43}
$$

where $\mathcal{T}$ represents the uncertainty term. And noting $n_1 = (y + L_p\sin\phi - y_{refp})\sin\phi, n_2 = (x + L_p\cos\phi - x_{refp})\cos\phi$, we have

$$
|\mathcal{T}| \le |F_1| + |(L_p - n_1 - n_2)F_2| = |F_1| + |(y - y_{refp})\sin\phi + (x - x_{refp})\cos\phi||F_2|.
$$

Since

$$|(y - y_{refp}) \sin \phi + (x - x_{refp}) \cos \phi| \leq |y - y_{refp}| + |x - x_{refp}|.$$

Finally,

$$\mathcal{T} \leq |F_1| + \left(|y - y_{refp}| + |x - x_{refp}|\right) |F_2| = |\bar{\mathcal{T}}|. \tag{4.44}$$

Thus,

$$\dot{V}_1 \leq S(\mathcal{M} + \mathcal{N}\delta_{sw} + \mathcal{R} + \bar{\mathcal{T}} + k\dot{e}_{yp}). \tag{4.45}$$

Although $\mathcal{T} \leq |\bar{\mathcal{T}}|$, it is not easy to determine the upper bound of $\bar{\mathcal{T}}$ in real applications; It is necessary to develop an adaptive controller for it. Assuming that $\bar{\mathcal{T}}$ is changing slowly, that means $\dot{\bar{\mathcal{T}}} = 0$, define a second Lyapunov function as

$$V_2 = V_1 + \frac{1}{2\lambda}\tilde{\mathcal{T}}^2, \tag{4.46}$$

where $\lambda > 0$, $\tilde{\mathcal{T}} = \hat{\mathcal{T}} - \bar{\mathcal{T}}$ is the error between $\bar{\mathcal{T}}$ and its estimation $\hat{\mathcal{T}}$.

Since assuming $\dot{\bar{\mathcal{T}}} = 0$,

$$\dot{V}_2 = \dot{V}_1 + \frac{1}{\lambda}\tilde{\mathcal{T}}\dot{\tilde{\mathcal{T}}} \leq S(\mathcal{M} + \mathcal{N}\delta_{sw} + \mathcal{R} + \bar{\mathcal{T}} + k\dot{e}_{yp}) + \frac{1}{\lambda}\tilde{\mathcal{T}}\dot{\tilde{\mathcal{T}}}, \tag{4.47}$$

set the control law as

$$\delta_{sw} = \frac{-\mathcal{M} - \mathcal{R} - \bar{\mathcal{T}} - k\dot{e}_{yp} - \epsilon \, \text{sign}(S)}{\mathcal{N}}. \tag{4.48}$$

Substitute (4.48) into (4.47), then

$$\dot{V}_2 \leq \hat{\mathcal{T}}(-S + \frac{1}{\lambda}\dot{\hat{\mathcal{T}}}) - \epsilon \, \text{sign}(S)S. \tag{4.49}$$

Therefore, we set the adaptive control law as

$$\dot{\hat{\mathcal{T}}} = S\lambda \ (\lambda > 0). \tag{4.50}$$

Again, the discontinuous component $\text{sign}(S)$ is replaced by a continuous function $\Pi(S)$ to eliminate the chattering effect, namely,

$$\Pi : S \rightarrow \frac{S}{|S| + \sigma}.$$

The completed steering control law (4.48) rewrites as follows:

$$\delta_{sw} = \frac{-\mathcal{M} - \mathcal{R} - \hat{\mathcal{T}} - k\dot{e}_{yp} - \epsilon\frac{S}{|S| + \sigma}}{\mathcal{N}}, \tag{4.51}$$

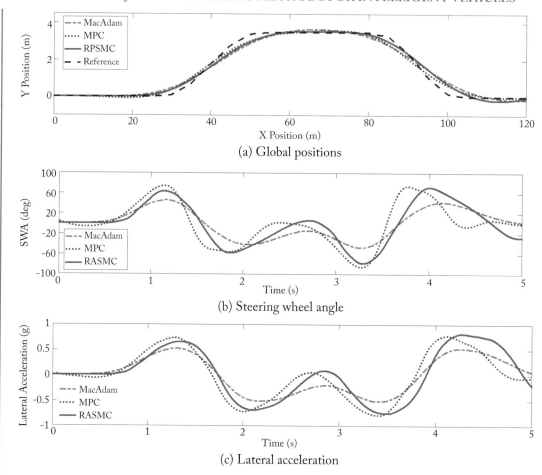

Figure 4.4: Simulation results of the vehicle with different steering control methods.

where $\hat{\mathcal{T}}$ is subjecting to the dynamical system (4.50). Finally, substituting (4.51) into (4.49), we have

$$\dot{V}_2 \leq S\left(-\tilde{\mathcal{T}} - \epsilon \frac{S}{|S| + \sigma}\right) - \frac{1}{\lambda}\tilde{\mathcal{T}}\dot{\tilde{\mathcal{T}}} = \tilde{\mathcal{T}}\left(-S + \frac{1}{\lambda}\dot{\tilde{\mathcal{T}}}\right) - \frac{\epsilon S^2}{|S| + \sigma} = -\frac{\epsilon S^2}{|S| + \sigma} < 0 \quad (4.52)$$

which is held for any $\epsilon > 0, \sigma > 0$, which means (4.51) could stabilize the path tracking system.

To show the effectiveness of the proposed adaptive robust SMC (RASMC) steering controller, the other two well-known driver models such as the optimal preview control model by MacAdam, and the conventional MPC method are compared. The test scenario is a Double Lane Change maneuver where the vehicle moves at a relatively high speed as 90 km/h. As shown in Figure 4.4a, we could observe that these entire three controllers could follow the target path well. Figure 4.4b also shows the steering wheel angle of the vehicle and Figure 4.4c shows the corresponding lateral accelerations. Overall, it seems that the tracking performance of the MPC controller and the RASMC controller is better than the preview method based on the optimal control, however, lateral accelerations of these two are relatively higher. Further, the change rate of the steering wheel angle of the RASMC is better than that of the MPC, which means it would be easier to manipulate the vehicle for the RASMC. Additionally, due to the mechanism of the controller design, the RASMC will not suffer from negative influences caused by vehicle system parameter variations, while the MacAdam model is still sensitive to the parameters of the vehicle (e.g., mass, tire cornering stiffness). Therefore, the proposed controller could track the target path well and robustly; It also maintains an easy operating performance.

4.3   TRAJECTORY TRACKING BY THE LINEARIZED TIME-VARYING MPC

In this section, we would design a trajectory-following controller based on the model predictive control approach. MPC is a well-known, effective control technical which is achieved by optimizing a finite time-horizon while only implementing the current timeslot to the control plant. It owns the ability to anticipate future events and take control actions accordingly. However, handling a nonlinear system via the nonlinear MPC is still challenging and time consuming, thus, a detailed linearized time-varying MPC controller derived from a nonlinear 5DOF vehicle model is presented to track the target planned path and velocity simultaneously in this section.

4.3.1   NONLINEAR 5DOF VEHICLE MODEL AND TIRE MODEL

Generally speaking, as shown in Figure 4.3a, the driver would process the road information from the perspective of the driver while following the road, which would be processed in the local driver coordinate system. Then a preview point (x_p, y_p) in the world coordinate system could be transformed into the local driver coordinate system (x', y') by

$$\begin{bmatrix} x' \\ y' \end{bmatrix} = \begin{bmatrix} \cos \psi & \sin \psi \\ -\sin \psi & \cos \psi \end{bmatrix} \begin{bmatrix} x_p - x_v \\ y_p - y_v \end{bmatrix}, \tag{4.53}$$

where ψ denotes the yaw angle of the host vehicle and (x_v, y_v) represents the coordinate of the vehicle's CoG in the world coordinate system.

As depicted in Figure 4.3b, if the roll motion of the front/rear wheels ω_f, ω_r are concluded, then a 5DOF single-track rigid body model that adds DOF of the vehicle longitudinal motion u, lateral motion v, and the yaw motion ψ can be derived and will be considered for prediction in the MPC algorithm. According to Newtown's second law, the motion of a vehicle is governed by the following nonlinear dynamical system:

$$\begin{cases} \dot{u} = \dfrac{1}{m}(F_{xwf} + F_{xwr}) + vr \\ \dot{\beta} = \dfrac{1}{m}(F_{ywf} + F_{ywr}) - r \\ \dot{\psi} = r \\ \dot{r} = \dfrac{1}{I_z}(aF_{ywf} - bF_{ywr}) \\ \dot{\omega}_f = \dfrac{1}{J_{wf}}(T_{wf} - F_{xwf}r_w) \\ \dot{\omega}_r = \dfrac{1}{J_{wr}}(T_{wr} - F_{xwr}r_w), \end{cases} \tag{4.54}$$

where m denotes the vehicle has a lumped mass, I_z denotes the moment of inertia respecting the Z-axis of the vehicle, F_{ywf}, F_{ywr} denote the lateral force of the front wheels and rear wheels, respectively, and F_{xwf}, F_{xwr} denote the longitudinal tire force of front wheels and rear wheels, respectively. The length from the front/rear axle to the vehicle CoG is denoted by a and b, respectively, J_{wf} and J_{wf} denotes the equivalent inertia of the front wheel and rear wheel, respectively, and r_w denotes the effective rolling radius of the wheels. T_{wf}, T_{wf} denote the torque acting on the front wheels and rear wheels, respectively, which could be a driving torque when in accelerating or a braking torque while in braking, for a rear-drive vehicle, which is,

$$T_{wf} = \begin{cases} \tau T_w & \text{if braking} \\ 0 & \text{if accelerating,} \end{cases} \qquad T_{wr} = \begin{cases} (1-\tau)T_w & \text{if braking} \\ T_w & \text{if accelerating,} \end{cases} \tag{4.55}$$

where $0 \leq \tau \leq 0$ denotes the braking torque distribution factor between the front wheels and rear wheels.

When in a working condition with a large lateral acceleration, the vehicle demands more lateral force from tires, however, the system would become highly nonlinear since the main cause of this nonlinearity comes from a limited available tire lateral force afforded by the road surface. One common way of modeling the tire force vs. slip is to use the analogous Magic Formula tire models under pure slip conditions, such that

$$\begin{cases} F_{xj}(\kappa_j) = \mu F_{zj} \sin\left(C_x \arctan(B_x \kappa_j / \mu)\right) \\ F_{yj}(\kappa_j) = \mu F_{zj} \sin\left(C_y \arctan(B_y \alpha_j / \mu)\right) \end{cases} \quad j \in \{f, r\}, \tag{4.56}$$

where $F_{z,j}$, $j \in \{f, r\}$ denote the vertical loads of the front wheel and rear wheel, respectively, and C_x, B_x, C_y, B_y are the parameters of the MF tire model. However, the vehicle might deal with a combined slip situation where the longitudinal speed would change, then the tire force is a function with longitudinal slip κ, lateral slip α, vertical loads F_z, and ground adhesion constant μ, namely,

$$\begin{cases} F_{xj}(\kappa_j) = f_x(\kappa_j, \alpha_j, F_{zf}, \mu) \\ F_{yj}(\kappa_j) = f_y(\kappa_j, \alpha_j, F_{zf}, \mu) \end{cases} \quad j \in \{f, r\}, \tag{4.57}$$

where κ denotes the longitudinal slip and α denotes the sideslip angle of the tire.

The vertical tire loads for the front tire and rear tire could be approximated by

$$\begin{cases} F_{zf} = \dfrac{mgb}{a+b} + \epsilon F_{drag} \\ F_{zr} = \dfrac{mga}{a+b} + (1-\epsilon)F_{drag}, \end{cases} \tag{4.58}$$

where ϵ denotes the wind drag-down force distribution factor, the wind drag-down force is calculated by $F_{drag} = \frac{1}{2}\rho C_z A_d u^2$, where ρ denotes the air density, C_z denotes the lift coefficient, and A_d denotes the frontal area of the vehicle.

Besides, let us define the subcomponents of the combined slip as

$$\begin{cases} \sigma_x = \dfrac{\kappa}{1+|\kappa|} \\ \sigma_y = \dfrac{\alpha}{1+|\kappa|}, \end{cases} \tag{4.59}$$

where κ denotes the tire longitudinal slip and α denotes the tire sideslip angle. Further, the tire sideslip angle α_f, α_r for front and rear wheels could be approximated by

$$\begin{cases} \alpha_f = \dfrac{\delta_{sw}}{G} - \dfrac{v+ar}{u} \\ \alpha_r = -\dfrac{v-br}{u}, \end{cases} \tag{4.60}$$

where the steering wheel angle is denoted by δ_{sw} through a steering gear ratio G to result in front-wheel steer angles.

Also, we define the longitudinal slip of the tire as

$$\kappa = \begin{cases} \dfrac{r_w\omega - u}{u} & \text{if } r_w\omega \le u \\ \dfrac{r_w\omega - u}{r_w\omega} & \text{otherwise}, \end{cases} \tag{4.61}$$

where ω denotes the wheel angular speed.

Then, the combined slip σ is evaluated by

$$\sigma = \sqrt{\sigma_x^2 + \sigma_y^2}. \tag{4.62}$$

Still, we need derive the tire force (4.57) based on the pure longitudinal/lateral tire forces (4.56), and if we want to use the combined slip quantities σ, we must have the pure slip characteristics with σ_x and σ_y as abscissa available. This could be simply realized by following relationships, which is

$$\begin{cases} F_{\sigma xj}(\sigma_x) = F_{xj}\left(\dfrac{\sigma_x}{1-|\sigma_x|}\right) \\ F_{\sigma yj}(\sigma_y) = F_{yj}\left(\arctan\left(\dfrac{\sigma_y}{1-|\sigma_x|}\right)\right) \end{cases}, \quad j \in \{f,r\}. \tag{4.63}$$

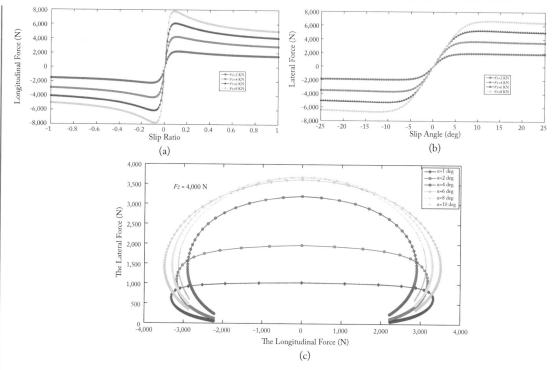

Figure 4.5: Tyre lateral and longitudinal characteristic curves: (a) longitudinal tire force, (b) lateral tire force, and (c) the relationship curve between the longitudinal and lateral forces.

Finally, the longitudinal force and the lateral force with combined slip are calculated by,

$$\begin{cases} F_{xwj}(\kappa_i, \alpha_i) = \dfrac{\sigma_{xj}}{\sigma_j} F_{\sigma xj}(\sigma_j) \\ F_{ywj}(\kappa_i, \alpha_i) = \dfrac{\sigma_{yj}}{\sigma_j} F_{\sigma yj}(\sigma_j) \end{cases}, \quad j \in \{f, r\}. \tag{4.64}$$

The lateral and longitudinal characteristic curves are shown in Figure 4.5.

4.3.2   MODEL LINEARIZATION AND OPTIMIZATION

To deploy the MPC method, at first, we have to reform (4.54) into a discrete form with a sample time T_s. With an arbitrary working point (x_0, u_0), the nonlinear vehicle dynamic model could be linearized as a discrete linear time-varying state-space form by taking the first-order term of its Taylor expansion, such that

$$\begin{cases} x(k+1) = f_v(x_0(k), u_0(k)) + \mathbf{A}(k)(x(k) - x_0(k)) + \mathbf{B}(u(k) - u_0(k)) \\ z(k) = g_v(x_0(k), u_0(k)) + \mathbf{C}(k)(x(k) - x_0(k)), \end{cases} \tag{4.65}$$

where the state variables $x = [u, v, \psi, r, \omega_f, \omega_r]^T$, input variables $u = [\delta_{sw}, T_w]^T$, and the outputs $z = [y, u]^T$, which includes the lateral position y and the vehicle longitudinal velocity u, $\mathbf{A}, \mathbf{B}, \mathbf{C}$ denote the Jacobian matrixes evaluated at the working point. Then the predicted outputs Z respecting to the dynamical system (4.65) could be calculated sequentially using the set of future control vector U and its current states, namely,

$$Z = \mathbf{F} x(k) + \mathbf{\Phi} U, \tag{4.66}$$

where

$$Z(k) = \begin{bmatrix} z(k) \\ z(k+1) \\ \vdots \\ z(k+N_p) \end{bmatrix} \quad U(k) = \begin{bmatrix} u(k) \\ u(k+1) \\ \vdots \\ u(k+N_p) \end{bmatrix} \quad \mathbf{F} = \begin{bmatrix} \mathbf{CA} \\ \mathbf{CA}^2 \\ \vdots \\ \mathbf{CA}^{N_p} \end{bmatrix}$$

$$\mathbf{\Phi} = \begin{bmatrix} \mathbf{CB} & 0 & \cdots & 0 \\ \mathbf{CAB} & \mathbf{CB} & \cdots & 0 \\ \vdots & \vdots & \ddots & \vdots \\ \mathbf{CA}^{N_p-1}\mathbf{B} & \mathbf{CA}^{N_p-2}\mathbf{B} & \cdots & \mathbf{CA}^{N_p-N_c}\mathbf{B} \end{bmatrix}$$

and N_p, N_c are the predictive horizon and control horizon, respectively.

Recalling the objective of the trajectory following is to track the planned path precisely, meanwhile, also keep the vehicle running at the desired speed. That means, for a given reference signal $r(k)$ at one instant k, the objective of the MPC is to ensure the predicted outputs as precisely close to the reference signal within the prediction horizon. Then the cost quadratic function is defined by

$$\tilde{J} = \sum_{i=1}^{N_p} ||\mathbf{\Pi}(k,k)(Y_{ref}(k+i|k) - Z(k+i|k))||^2 + \sum_{i=0}^{N_c-1} ||\mathbf{\Omega}(k,k)U(k+i|k)||^2, \tag{4.67}$$

where $\mathbf{\Pi}(k,k)$ denotes a diagonal subcomponent weight matrix for the errors between the reference signals and predicted outputs, Y_{ref} denotes the reference signal including the vehicle's lateral position and its longitudinal speed, $\mathbf{\Omega}(k,k)$ denotes the weight matrix component for the predicted control input. In essence, the optimal problem for MPC could be converted to a standard quadratic programming problem with the decision variable denoted by U, which is,

$$\min \tilde{J} = \frac{1}{2}U^T \widetilde{H} U + \widetilde{G}^T U, \tag{4.68}$$

where $\tilde{\mathbf{H}}, \tilde{\mathbf{G}}$ are compatible matrices and vectors in the quadratic programming problem. In our case, those matrices are

$$\tilde{\mathbf{H}} = 2(\mathbf{\Omega} + \mathbf{\Phi}^T \mathbf{\Pi} \mathbf{\Phi}), \tilde{\mathbf{G}} = (2\mathbf{\Lambda}^T \mathbf{\Pi} \mathbf{\Phi})^T, \mathbf{\Lambda} = \mathbf{F} x(k) - Y_{ref}. \tag{4.69}$$

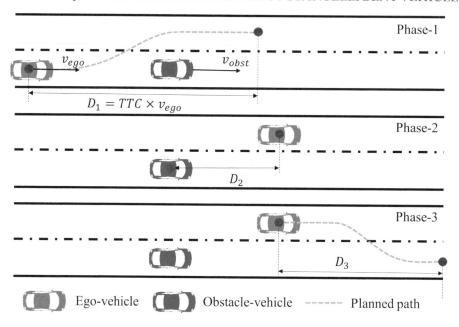

Figure 4.6: Procedures of overtaking maneuver.

The input constraints are expressed in terms of U, as well as the incremental input constraints, are

$$U_{\min} \leqslant u \leqslant U_{\max}, \quad \Delta U_{\min} \leqslant \Delta u \leqslant \Delta U_{\max}. \tag{4.70}$$

The solution for the quadratic program problem could be solved by *quadprog* in Matlab, and the active-set algorithm is chosen in this short-book. Finally, once the optimal control inputs are calculated, the driving torque acting on the wheels should be converted to the engine throttle position γ through the powertrain system as shown in (4.27).

4.3.3   CASE STUDY

In this section, we will focus on the demonstration on overtaking maneuver where the trajectory profiles are tracked by the linearized MPC controller, as introduced in this chapter, whose parameter settings are elaborated in Table 4.1. Overtaking is the act of the ego vehicle passing the slower-moving vehicle ahead on the road while traveling in the same direction. As depicted in Figure 4.6, a completed overtaking maneuver includes three sub-phases.

(1) **Phase 1 (Lane changing):** If an obstacle vehicle is detected ahead in the same lane, and the safe criteria for lane changing are met (e.g., Time_headway>2.5 s), then the trajectory-planning module would be triggered and generate a feasible path and velocity profiles for lane changing by the convex optimization and natural cubic splines

method described in Chapters 4 and 5. The planning distance $D_1 = TTC v_{ego}$, where TTC denotes the current Time-to-Collision, v_{ego} denotes the speed of the ego vehicle, then the ego vehicle shall follow the planned path and planned speed by the linearized MPC to move to the center of the adjacent lane.

(2) **Phase 2 (Lane keeping until passing the obstacle):** After the vehicle reaches the destination again, if no obstacle is detected ahead in the current lane of the time headway is larger than 5 s, then the vehicle would keep its lane to pass the obstacle vehicle with a safe distance D_2, while the target path and speed profiles for lane keeping are obtained by the convex optimization and natural cubic splines method described in Chapters 3 and 4, and the trajectory profiles are tracked by the linearized MPC method.

(3) **Phase 3 (Returning back-R.B.):** Similar to Phase 1, the ego vehicle makes a lane change with a planning distance D_3 returning to its original lane to complete the overtaking maneuver.

Interacting with a Moving Vehicle on a Straight Road

At first, let us take a look at the overtaking maneuver on a straight road. As shown in Figure 4.7a, the total length of the straight road segment is around 450 m, and there is an obstacle vehicle ahead in the right lane with a constant moving speed as 15 m/s. Besides, the initial speed of the host vehicle is 20 m/s, and their distance gap is 30 m at the very beginning. The driving task demands that the ego vehicle safely overtakes the obstacle vehicle. The maximum speed is set to be less than 25 m/s, and the expected lateral accelerations are less than 4 m/s².

The global following path of the ego vehicle and obstacle vehicle are shown in Figure 4.7a. At first, the adaptive following module is activated since the time headway for lane changing is not enough. Once the criteria for a lane changing are met (time_headway>2.5 s), the host vehicle would perform a left lane change maneuver to its adjacent lane. Then it would keep in the left lane until the ego vehicle passes the obstacle vehicle by a safe distance ($D_2 = 20$ m). Finally, the ego vehicle would return to the original right lane to complete the overtaking maneuver.

Besides, the followed speed is shown in Figure 4.7b, which exhibits that the ego vehicle would slow down its speed to around 15 m/s to increase the time headway to satisfy the lane-changing criteria. Then the ego vehicle would gradually speed up to the maximum allowed speed until the overtaking maneuver is finished. The control inputs throttle, braking torques, and the steering wheel angle) for the ego vehicle as shown in Figures 4.7c,d are smooth as well. While according to Figures 4.7d,e, the maximum absolute value of the yaw rate and the lateral acceleration is around 4°/s and 1.8 m/s², respectively, which means the vehicle will move stably. As for the driving comfort as indicated in Figure 4.7e, the longitudinal acceleration of the ego vehicle lies in the acceptable interval of -2.9 m/s² to 1.0 m/s².

Table 4.1: Parameters of the ego vehicle and the MPC controller

Parameter Description	Values
Longitudinal MF constants B_x, C_x	16.480 [-], 1.217 [-]
Lateral MF constants B_y, C_y	-11.270 [-],1.452 [-]
Steering ratio G	18 [-]
Vehicle mass m, γ	1800 kg
Vehicle inertia respecting to Z-axis I_z	3200 kg $\cdot$ m^2
TWheel inertia J_{wf}, J_{wr}	2.9 kg $\cdot$ m^2
The distance from the CoG to the front axle a	1.4 m
The distance from the CoG to the front axle b	1.65 m
Wheel effective radius r_w	0.36 m
The gear ratio of the differential i_g	4.1 [-]
The efficiency of the powertrain η	0.95 [-]
Braking distribution factor τ	0.5 [-]
Wind drag-down force distribution ϵ	0.4 [-]
Frontal area A_d	2.2 m^2
Air density ρ	1.206 kg/m^3
Lift coefficient	0.18 [-]
Gravitational constant g	9.8 m/s^2
Road adhesion constant μ	0.9 [-]
MPC discretization sample time T_s	0.02 s
Minimum constraints of incremental input ΔU_{min}	[-15 deg, -500 N $\cdot$ m]
Maximum constraints of incremental input ΔU_{max}	[15 deg, 300 N $\cdot$ m]
Minimum constraints of input U_{min}	[-150 deg, -3000 N $\cdot$ m]
Maximum constraints of input U_{max}	[150 deg, 1600 Nm]
Prediction/control horizontal length N_p, N_c	50 [-], 50 [-]

Interacting with a Moving Vehicle on a Curved Road

In this section, we will test if the presented framework would be effective as well when the ego vehicle performs an overtaking maneuver on a curved road. As shown in Figure 4.8a, the total length of the curved road is around 550 m. Similar to the previous scenario, there is also an obstacle vehicle ahead in the right lane with a constant moving speed as 15 m/s, and their distance gap is 30 m. The initial speed of the ego vehicle is 20 m/s, and the driving task demands the

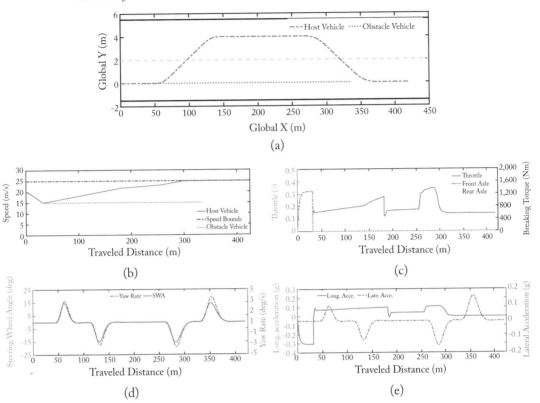

Figure 4.7: The trajectory following results of overtaking on a straight road.

ego vehicle safely overtake the obstacle vehicle as quickly as it can. Again, for safety considerations on a curved road, the maximum speed should be less than 20 m/s, and the expected lateral accelerations are less than 4 m/s².

The global following path of the ego vehicle and obstacle vehicle are shown in Figure 4.8a; the overtaking procedure is quite similar to the previous one. What is more, the actual followed traveling speed as depicted in Figure 4.8d shows that the ego vehicle would slow down its speed to around 13 m/s to satisfy the lane-changing criteria. Then it would gradually accelerate to the maximum allowed speed (20 m/s) after reaching the adjacent lane. Besides, as shown in Figures 4.8c,d, the inputs (throttle and braking torques and the steering wheel angle) for the ego vehicle are smooth as well. The maximum absolute value of the yaw rate and the lateral acceleration is around 6°/s and 2 m/s², respectively, as shown in Figures 4.8d,e, which indicate the vehicle moves in an excellent stable condition on a curved road. Figure 4.8e also shows that the longitudinal acceleration of the ego vehicle lies in the acceptable interval of -2.9 m/s² to 1.2 m/s². Therefore, the framework shall work on a curved road.

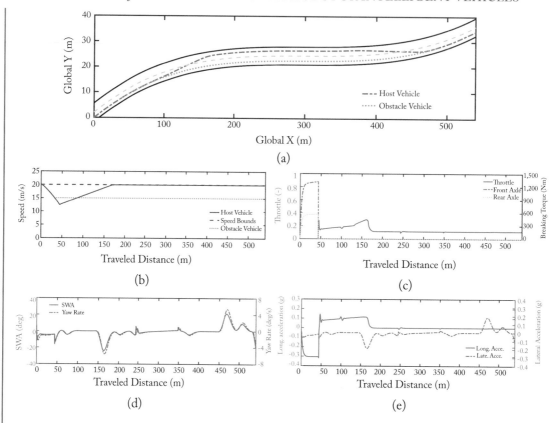

Figure 4.8: The trajectory following results of overtaking on a curved road.

4.4   SUMMARY

We mainly introduce robust trajectory-following methods for intelligent vehicles by the adaptive SMC which needs to be designed separately for speed tracking or path tracking, while the linearized time-varying MPC based on a 5DOF nonlinear vehicle model can be utilized for the tracking of the speed and the path simultaneously. Both methods can be insensitive to model uncertainties and disturbances, such that they will work well for different driving scenarios. Due to the well-studied convex optimization, the solution can be solved in the scale of milliseconds, which is promising for the onboard implementation on the intelligent vehicle, and it will be included in our further work.

C H A P T E R   5

Control Strategies for Human-Automation Cooperative Driving Systems

In this chapter, we will introduce our research work on control strategies for human-automation cooperative driving system. The driving intention, driving risk, and the performance index of the control system will be considered for the design of the cooperative driving strategy. The driving intention experiment and two machine learning methods (e.g., SVM and iMLCU) are introduced to recognize the driver's driving intention. Driving safety field and TTC-based evaluation methods are adopted to assess the driving risk. There are two cooperative driving strategies based on the game theory and fuzzy control theory will be introduced to realize human-automation cooperative driving and help the vehicle safely avoid obstacles.

5.1   DRIVING INTENTION RECOGNITION

In this section, the driver-in-the-loop experiment for driving intention data collection will be introduced, including the experiment scenario, data processing, and labeling. To recognize the driving intention, two machine learning methods, including the SVM and iMLCU approach, are applied for the driving intention recognition model.

5.1.1   DATA COLLECTING EXPERIMENT AND PROCESSING

We will introduce experiment details for driving intention data-collecting, including experiment equipment, the designed scenario, data collecting procedure, and data processing. The collected data are used for driving intention recognition training and testing.

Equipment and Scenario

The driving intention data collecting experiment was designed based on the driver-in-the-loop simulation platform (shown in Figure 5.1). Logitech G27 was used for the human driver's control action signals collection, including steering wheel angle, throttle opening, and braking pedal displacement in real time, which was supported via the *CarSim-LabVIEW* simulation environment. Moreover, the data-collecting scenario considering different road shapes and obstacle vehicles is designed by *CarSim* software. As illustrated in Figure 5.2, the driving scenario con-

Figure 5.1: Driver-in-the-loop simulation platform.

sists of four segments, including two straight road segments, a positive curvature, and a negative curvature curvy road. Besides, there are four obstacles with different initial states. The details of each obstacle vehicle are described below.

(1) O_1 is a fixed obstacle vehicle with its initial station at 120 m.

(2) O_2 is a dynamic obstacle vehicle running along a center line of right lane with the initial station at 200 m with constant velocity at 10 m/s.

(3) O_3 is a dynamic obstacle vehicle running along center line of right lane with the initial station at 650 m with constant velocity at 5 m/s.

(4) O_4 is a fixed obstacle vehicle with its initial station at 1,350 m.

Task and Procedure

The primary task of this driving intention data-collecting experiment is to avoid obstacles by lane-changing maneuvers, and the participant should turn back to the original lane after exceeding the obstacle vehicle to increase the data collecting efficiency. During the whole data collecting experiment process, every participant should keep the following rules.

(1) Every participant should be under normal physical and psychological states. Fatigue and distracted driving states are not allowed.

(2) Smartphones are not allowed.

(3) Vehicle velocity should be not larger than 120 km/h.

(4) Each participant is allowed to be familiar with the platform and the scenario by test driving two times.

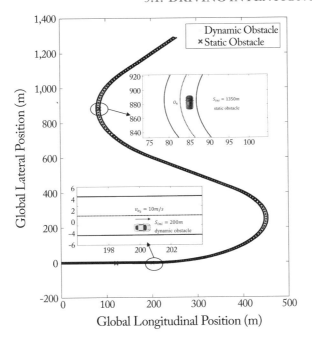

Figure 5.2: Illustration of data collecting experiment scenario.

Data Collecting and Labeling

The driving intention is divided into three categories, including the left lane-changing (LLC), lane-keeping (LK), and right lane-changing (RLC). To investigate the driving intention, three characteristic features are selected, including the steering wheel angle (SWA), the relative lateral offset to the road center line y_{rc}, and relative lateral velocity v_{yrc}. SWA is an important characteristic of lane-changing maneuver [96], thus, it should be taken into consideration. The other two characteristics, y_{rc} and v_{yrc}, could reflect vehicle lane-changing relative motion features, which are determined by Equation (5.1):

$$y_{rc} = y - r_c, \ v_{y_{rc}} = \dot{y} - \dot{r}_c,$$

(5.1)

where r_c is the lateral position of the road center line. As shown in Figure 5.4, when the human driver keeps current lane, the value of v_{yrc} holds around 0, and the value of y_{rc} is about the half of lane width. Moreover, when the human driver executes the lane-changing maneuver, the value of v_{yrc} grows and the value of y_{rc} is varied between the two road lanes.

Data labeling has influences on the performance of the supervised machine learning or semi-supervised methods, and the distinct data labeling principle is required. As shown in Figure 5.3, there are obvious data features for three lane-changing maneuvers labeling. For example, both LLC and RLC lane-changing categories have a absolute larger value of v_{yrc} than LK (e.g.,

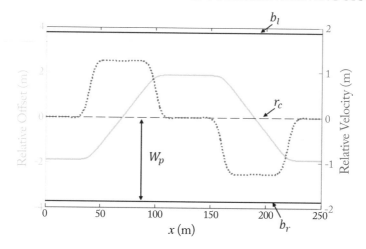

Figure 5.3: Illustration of lane-changing maneuver feature selection.

$v_{yrc} \geq 1.8$ m/s). For the category of LK, the values of y_{rc} are centralized around the half of lane width and v_{yrc} nears around 0.

5.1.2   SUPPORT VECTOR MACHINE

SVM is a supervised machine learning method with an associated learning algorithm and it can achieve great classification performance. Therefore, SVM is used to recognize driver's driving intention and the SVM algorithm will be introduced briefly in the following part.

As shown in Figure 5.4, given a set of training pairs $\{x_i, y_i\}$, each training data x_i marked as possible group labels $y_j \in \{-1, 1\}$, an SVM training algorithm builds a model to find the hyperplane that is used to assign a new example to one group or the other group. The highest classification accuracy is achieved by finding the hyperplane that has the largest distance to the nearest training data point. Moreover, the hyperplane is defined as

$$\boldsymbol{\omega} x + b = 0, \tag{5.2}$$

where $\boldsymbol{\omega}$ means the normal vector to the hyperplane and b denotes bias. Furthermore, the parameter $\frac{b}{||\boldsymbol{\omega}||}$ determines the offset between the hyperplane and the origin along the direction of $\boldsymbol{\omega}$. If the training data is linearly separable, the two parallel hyperplanes could be selected, defined as

$$\begin{cases} \boldsymbol{\omega} x + b = 1 \\ \boldsymbol{\omega} x + b = -1. \end{cases} \tag{5.3}$$

Generally, the distance between these two hyperplane is $\frac{2}{||\boldsymbol{\omega}||}$. If the training data is nonlinearly separable, the kernel function is used to map the training data to a higher dimensional space to

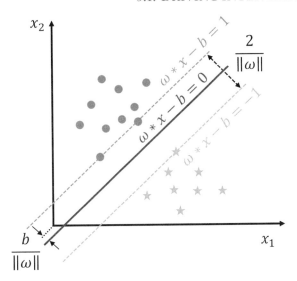

Figure 5.4: Illustration of SVM and its maximum-margin hyperplane.

find better hyperplane, and it is denoted as

$$k(\boldsymbol{x}_i, \boldsymbol{x}_j) = \phi(\boldsymbol{x}_i, \boldsymbol{x}_j), \tag{5.4}$$

where $\phi(\boldsymbol{x})$ is the feature map. In general, the radial basis function (RBF) is used as the kernel function,

$$k(\boldsymbol{x}_i, \boldsymbol{x}_j) = \exp\left(-\gamma \|\boldsymbol{x}_i - \boldsymbol{x}_j\|^2\right), \tag{5.5}$$

where γ is the Gaussian kernel parameter. Therefore, the best hyperplane is optimized from the following defined problem in the SVM model:

$$\min_{\boldsymbol{w}, b, \zeta} \frac{1}{2} \boldsymbol{w}^T \boldsymbol{w} + C \sum_{i=1}^{N} \zeta_i \tag{5.6}$$
$$\text{s.t. } y_i(\boldsymbol{w}^T \phi(\boldsymbol{x}_i) + b) \geq 1 - \zeta_i, \ \zeta_i \geq 0,$$

where C means the penalty parameter and ζ_i refers to the relaxation factors. Equation (5.6) is a convex optimization problem that can be optimized by transforming into the Lagrangian dual form, namely,

$$\max_{\alpha} \sum_{i=1}^{N} \alpha_i - \frac{1}{2} \|f\|_H^2 \tag{5.7}$$
$$\text{s.t. } \sum_{i=1}^{N} y_i \alpha_i = 0, \ 0 \leq \alpha_i \leq C, i = 1, \ldots, N$$

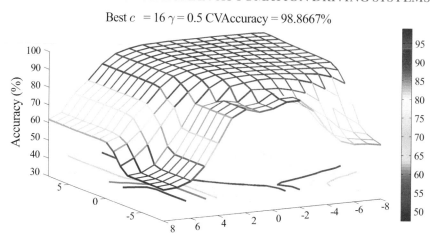

Figure 5.5: The CV result of the parameter C and γ.

with

$$\|f\|_H^2 = \sum_{i=1}^{N} \sum_{j=1}^{N} y_i \alpha_i (\phi(x_i) \cdot \phi(x_j)) \alpha_j y_j.$$

The Lagrangian dual problem can be well solved by using the quadratic programming (QP) algorithm. After obtaining the optimized parameter α_i, the best hyperplane can be determined by

$$\omega = \sum_{i=1}^{N} \alpha_i y_i \phi(x), \ b_i = \omega \phi(x) - y_i. \tag{5.8}$$

The solution of the optimized problem (5.6)–(5.8) can be well solved by a toolbox of LIBSVM in MATLAB environment. To avoid data over-fitting problem and obtain better classification accuracy, the cross validation (CV) is required to search best model parameters C and γ. The k-fold method is used to search the combination of C and γ, and the best parameter will be determined according to the highest CV accuracy. As shown in Figure 5.5, the best combination of $C = 16$, and $\gamma = 0.5$ with 99.87% accuracy are selected for the SVM model.

Testing Results

To validate the SVM model for the driver's driving intention recognition, a testing dataset and testing scenario are provided. The testing dataset consists of 200 groups data of each driving intention category, 600 groups in total. The testing results are listed in Table 5.1. The average recognition accuracy of the SVM model is 97.8%, and the accuracy of each category is higher than 96%.

Table 5.1: The accuracy of the SVM model

	LLC	LK	RLC	Accuracy
LLC	194	6	0	97
LK	0	200	0	100
RLC	0	7	193	96.5
Average				97.8

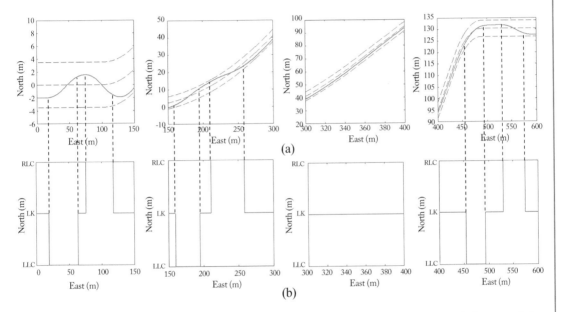

Figure 5.6: The SVM model recognition result in the design scenario: (a) the road shape and the vehicle trajectory and (b) the lane-changing intention recognition result.

The testing scenario consists of two straight roads, a positive curvature, and a negative curvy road. There are overtaking maneuver (including LLC and RLC actions) in the first, second, and last subfigures, and lane-keeping maneuver in the third subfigure, as depicted in Figure 5.6a. The result of driving intention recognition is shown in Figure 5.6b. The dashed line connects the vehicle trajectory with driving intention, which indicates that the recognized driving intention can reflect driver's driving intention and action changes.

5.1.3   iMLCU METHOD

An inductive multi-label classification with an unlabeled data (iMLCU) method is proposed to address the semi-supervised multiple-label learning problem. Compared with the SVM method, the training dataset comprises the labeled training dataset S^l and the unlabeled training dataset S^u. Given a set of labeled training dataset $S^l = \{(\mathbf{x}_i, Y_i)\}$, $\mathbf{x}_i \in R^d$ are the d-dimensional training data point, $Y_i = \{y_1, \ldots, y_l\}$ are the label space of $\mathbf{x}_i$ with q possible classes and $y_j \in \{-1, 1\}, 1 \leq j \leq q$. Supposing that there are n labeled training instances and m unlabeled training instances, therefore the training dataset S is denoted as

$$S = \{(\mathbf{x}_1, Y_1), \ldots, (\mathbf{x}_n, Y_n), \mathbf{x}_{n+1}, \ldots, \mathbf{x}_{n+m}\}. \tag{5.9}$$

The goal of iMLCU classifier is to find a family of q real-value functions $f_i(\mathbf{x}, y_i)$ to predict a correct label for the new test data S^t, and $f_i(\mathbf{x}, y_i)$ are given by

$$f_i(\mathbf{x}, y_i) = \langle \mathbf{w}_i, \mathbf{x} \rangle + b_i, 1 \leq i \leq q, \tag{5.10}$$

where $\langle \mathbf{w}_i, \mathbf{x} \rangle$ is the inner product of the weight vector $\mathbf{w}_i$ and the training data $\mathbf{x}$, and b_i is the bias for class label y_i. Simultaneously, to address the multiple-label problem, the method to predict the label of test data is adopted as follows:

$$\hat{Y} = \text{sign}\,(f_1(\mathbf{x}, y_1), \ldots, f_q(\mathbf{x}, y_q)). \tag{5.11}$$

The iMLCU classifier is trained by using the information of labeled and unlabeled data. Therefore, the cost function of iMLCU classifier comprises two parts, namely, the labeled part and the unlabeled part. The details of the cost function are described as follows.

Labeled Part

The hyperplanes for different classes labeled instances $(\mathbf{x}_i, Y_i)$ are defined as

$$\langle \mathbf{w}_k - \mathbf{w}_n, \mathbf{x}_i \rangle + b_k - b_n = 0, \tag{5.12}$$

where $(y_k, y_n) \in Y_i \times \bar{Y}_i$. High classification accuracy is achieved by finding the hyperplane that has the maximum margin to the nearest data of any training classes. Therefore, the labeled data are used by optimizing the following equation:

$$\min_{\mathbf{w}, \Xi} \sum_{k=1}^{q} \|\mathbf{w}_k\|^2 + C \sum_{i=1}^{n} \frac{1}{|Y_i||\bar{Y}_i|} \sum_{(y_k, y_n) \in Y_i \times \bar{Y}_i} \xi_{ikn}$$
$$\text{s.t.} \langle \mathbf{w}_k - \mathbf{w}_n, \mathbf{x}_i \rangle + b_k - b_n \geq 1 - \xi_{ikn} \tag{5.13}$$
$$\xi_{ikn} \geq 0, \; (1 \leq i \leq n, (y_k, y_n) \in Y_i \times \bar{Y}_i),$$

where $\Xi = \{\xi_{ikn} | 1 \leq i \leq n, (y_k, y_n) \in Y_i \times \bar{Y}_i\}$ relate to the slack variables ξ_{ikn} and C refers to the penalty parameter.

Unlabeled Part

The unlabeled instances should be placed outside the margin and penalized when it lies within the margin or even on the wrong side of the hyperplane in the learning process. However, without knowing the class label of an unlabeled instance, it can't be determined whether the unlabeled instance is on the right or wrong side of the hyperplane. Therefore, the iMLCU classifier adopts the same idea of S3VM to the unlabeled instances. The prediction label obtained from (5.10) is treated as the putative label sets of an unlabeled instance $\mathbf{x}$. Besides, the loss on label y_i is penalized by using the hinge loss function on $\mathbf{x}$, denoted as

$$\ell_i(\mathbf{x}, \hat{y}_i, f_i(\mathbf{x}, y_i)) = \max(1 - |\langle \mathbf{w}_i, \mathbf{x} \rangle + b_i|, 0). \tag{5.14}$$

For the sake of better classification accuracy, the total losses on unlabeled instances should be minimized, given by

$$\min_{\mathbf{w}} \sum_{j=n+1}^{n+m} \sum_{v=1}^{q} \max(1 - |\langle \mathbf{w}_v, \mathbf{x}_j \rangle + b_v|, 0). \tag{5.15}$$

However, Equation (5.15) is non-convex objective function because it consists of the sum of q non-convex functions ℓ_i on every unlabeled instance. In order to solve the non-convex optimization problem presented in this paper, the ConCave Convex Procedure (CCCP) method [97] is adopted. Therefore, (5.15) should be decomposed into a convex part and concave part. If an unlabeled instance $\mathbf{x}_j$ is predicted as a positive label y_v at current step, the effective loss function at next iteration is rewritten as follows:

$$\widetilde{L}(t) = \begin{cases} 0 & t \geq 1 \\ 1 - t & |t| < 1 \\ -2t & t \leq -1 \end{cases} \tag{5.16}$$

with $t = \langle \mathbf{w}_v, \mathbf{x}_j \rangle + b_v$. And for the case of the negative predicted label y_v, the effective loss function is defined as

$$\widetilde{L}(t) = \begin{cases} 2t & t \geq 1 \\ 1 + t & |t| < 1 \\ 0 & t \leq -1. \end{cases} \tag{5.17}$$

Therefore, the objective function of iMLCU classifier for the semi-supervised multi-label learning problem is defined as

$$
\min_{\mathbf{w}, \Xi} \underbrace{\sum_{k=1}^{q} \|\mathbf{w}_k\|^2 + C_1 \sum_{i=1}^{n} \frac{1}{|Y_i||\bar{Y}_i|} \sum_{(y_k, y_n) \in Y_i \times \bar{Y}_i} \xi_{ikn}}_{\text{Labeled part}}
$$

$$
+ C_2 \underbrace{\sum_{j=n+1}^{n+m} \sum_{v=1}^{q} \widetilde{L}(\langle \mathbf{w}_v, \mathbf{x}_j \rangle + b_v)}_{\text{Unlabeled part}} \qquad (5.18)
$$

$$
\text{s.t. } \langle \mathbf{w}_k - \mathbf{w}_n, \mathbf{x}_i \rangle + b_k - b_n \geq 1 - \xi_{ikn}
$$

$$
\xi_{ikn} \geq 0, \ (1 \leq i \leq n, (y_k, y_n) \in Y_i \times \bar{Y}_i)
$$

$$
\frac{1}{m} \sum_{j=n+1}^{n+m} \langle \mathbf{w}_v, \mathbf{x}_j \rangle + b_v = \frac{1}{n} \sum_{i=1}^{n} y_{iv},
$$

where C_1 and C_2 are non-negative weighting constants between the labeled part and the unlabeled part, and the value of these two parameters are referred from [98].

Training and Testing Results

To obtain better recognition accuracy, the label ratio between the labeled dataset and the unlabeled dataset should be considered since it has influence on the iMLCU model accuracy and training time. Here, we studied influences on different training datasets with five groups label ratios, namely, 1%, 2%, 3%, 4%, and 5%. The training datasets of different label ratios are made up of the same amount of the unlabeled data and different number of the labeled data, which are used for iMLUC classifiers training. Besides, to make the labeled data of three driving intentions balanced, the equal labeled data are randomly picked up from three categories. As shown in Figure 5.7, the average accuracy of iMLCU classifier increases with the label ratio, and when the iMLCU classifier is trained with 5% label ration dataset, it obtains 93.18% accuracy performance. However, the time consumption is also increased with the label ratio [98]. Therefore, 5% label ratio is adopted to train the iMLCU model to recognize the driver's driving intention, which is also used to develop the human-automation cooperative driving system.

5.2   SITUATION ASSESSMENT

In this section, the situation assessment methods are introduced to evaluate the driving risk, which has great influence on the decision making and control authority between the human driver and the automated driving system. Driving safety field is used to assess the driving risk imposed from obstacles, road boundaries, and driver's driving behaviors. The TTC-based method is adopted to evaluate the collision risk between the ego vehicle and front obstacles. Both the

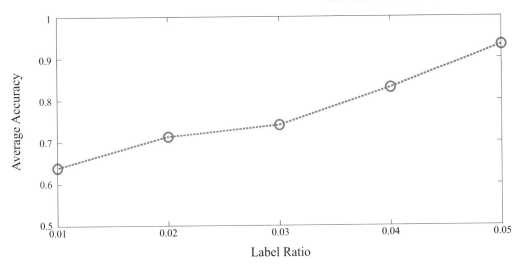

Figure 5.7: Accuracy of iMLCU classifiers which are trained with different label ratios.

driving safety field and TTC-based evaluation are used into the human-automation cooperative driving strategy.

5.2.1   DRIVING SAFETY FIELD

Driving safety field is developed based on the artificial field approach and it is used to assess the driving risk of the host vehicle, which takes the influences of the driver-vehicle-road interaction into consideration [99], including obstacles, road boundaries, driving behaviors, etc. In this part, the potential field, kinetic field, and behavior field will be introduced, respectively.

Potential Field

The potential field is an artificial field representing the non-moving objects on the driving safety, including two categories: obstacle potential and road potential. The objective of the obstacle potential is to evaluate collision possibilities and risks between the host vehicle and non-moving obstacles. The shorter relative distance between the vehicle and obstacles is, the higher collision risk would be and the incremental rate of collision possibilities should be decreased with the relative distance. Thus, the obstacle potential of the non-moving obstacle is denoted as

$$E_{R_o} = \frac{M_o R_o}{|r_o|} \frac{r_o}{|r_o|},\tag{5.19}$$

where M_o is the virtual mass of the non-moving obstacle that is related to the vehicle mass m_o and its velocity, R_o refers to the road condition influence factor related to road conditions [100]. $r_o = (x - x_o, y - y_o)$ means the distance vector, where (x, y) and (x_o, y_o) are the position vector of the host vehicle and the obstacle vehicle at the earth coordinate, respectively. In order to consider the differences between the tangent and normal direction on the driving safety, the enlarged distance vector r_o^* is introduced, defined as

$$r_o^* = \left(\frac{x^* - x_o^*}{a} \quad \frac{y^* - y_o^*}{b} \right),$$

(5.20)

where a and b are the scaling factors of the tangent and normal direction, which could be tuned according to environment, and they are determined by

$$a = |v_e - v_o| t_s, \; b = \tau,$$

(5.21)

where v_e and v_o refer to the longitudinal velocity of the host vehicle and the obstacle vehicle, respectively. t_s is the margin of TTC, and τ means the radius of safety circle. Obviously, the road shape is not considered in the above-mentioned obstacle potential. The coordinate-transition function between the earth coordinate and the local coordinate of the road center line is adopted to make the obstacle potential adapt to different road shapes, defined as

$$\begin{bmatrix} x^* \\ y^* \end{bmatrix} = \begin{bmatrix} \cos\theta & \sin\theta \\ -\sin\theta & \cos\theta \end{bmatrix} \begin{bmatrix} x \\ y \end{bmatrix},$$

(5.22)

where (x^*, y^*) represents the position vector at the local coordinate of the road center line and θ is the coordinate transition angle between these two coordinates.

The second category is the road potential, and its purpose is to prevent the vehicle violating road boundaries. The closer the position of the vehicle is to road boundaries, the higher possibilities of road departure and the traffic accident risk will be. The growing ratio of the road potential should be increased with the reduced relative distance between the vehicle to road boundaries. The road potential at (x_a, y_a) is expressed as

$$E_{Ra,q} = \eta_q R_a \ln \left(\frac{W}{2} - |r_a| \right) \frac{r_a}{|r_a|},$$

(5.23)

where η_q is scaling factors of the road boundaries, and $q \in \{u, l\}$ refer to the upper boundary and lower boundaries, respectively. W is the road width, and $r_a = (x - x_{a,q}, y - y_{a,q})$ denotes the distance vector from the vehicle to the road boundaries. The potential field with two non-moving obstacles located on different lanes and curvy road boundaries and the host vehicle moving at the constant speed of 20 m/s is shown in Figure 5.8.

Kinetic Field

The kinetic field is an artificial field denoting influences of moving objects on the driving risk. The relative movement characteristics including the relative distance and the movement direction are considered in the kinetic field when compared with the potential field. The driving risk reaches the highest value when the vehicle approaches from the front of the obstacle, and reaches the

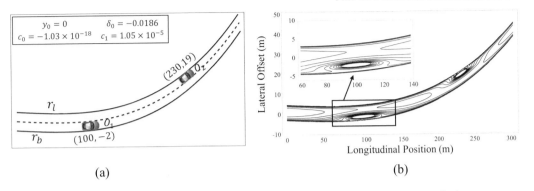

Figure 5.8: Illustration of the elliptic potential field: (a) designed scenario including two non-moving obstacles under curvy road boundaries and (b) distribution of the potential field.

lowest value when the vehicle approaches from the rear of the obstacle. As described above, the kinetic field at (x_v, y_v) is defined as

$$E_v = \frac{M_v R_v}{|r_v^*|} \frac{r_v}{|r_v|} \exp(k_1 v_v \cos \theta_v), \tag{5.24}$$

where r_v^* is the enlarged distance vector between the vehicle and the moving object with the transition coordinate matrix Equation (5.20). k_1 is the scaling constant that is greater than 0. θ_v denotes the direction angle between the vehicle and the velocity angle of the moving obstacle.

Behavior Field

The behavior filed is an artificial filed denoting the influence of the characteristics of driver behaviors. For instance, the driving risk provided by the behavior filed of an unskilled driver is always larger than that provided by the behavior field of a skilled driver, because the driving risk caused by the former is often higher than that caused by the latter. However, the driving risk caused by the behavior field to the surroundings is delivered by the moving vehicle driven by a driver. The behavior field of a driver could be denoted as the product of the kinetic field of the host vehicle and a driving risk factor at (x_c, y_c)

$$E_D = Df_c E_v, \tag{5.25}$$

where Df_c is the driving risk factor determined by the psychology and physiology, driving skill level, cognition, and traffic violations of a driver. E_v is the kinetic field of the host vehicle, and the direction of E_D is same as E_v. The larger value of E_D reflects the higher risk caused by the human driver's behavior.

The value of the driving safety field includes the evaluated value of the potential field, kinetic field, and behavior field, and it can be constructed as

$$E_s = E_p + E_v + E_D. \tag{5.26}$$

Table 5.2: Parameters of the driving safety field

Symbol	Description	Value	Units
m_o	Mass of obstacle vehicles	1,864	kg
R_i	Road condition influence factor	1	–
t_s	Margin of time to collision	2	s
τ	Radius of safety circle	2	m
η_q	Scaling factor of road border	-150	–
k_1	Scaling weight of kinetic field	0.005	–
k_2	Scaling weight of field force	0.005	–
Df_i	Driving risk factor of behavior field	0.5	–

The driving risk D_R will be evaluated by the field force imposed to the host vehicle, and it's calculated by

$$\boldsymbol{D}_{Ri} = \boldsymbol{E}_{si} M_i [R_i \exp(-k_2 u_i \cos \theta_i)(1 + Df_i)], \tag{5.27}$$

where k_2 refers to the scaling weight constant value, u_i is the velocity of the host vehicle and θ_i means the angle between the angle of $\boldsymbol{E}_{si}$ and the direction of u_i. The parameters of the driving safety field are listed in Table 5.2.

5.2.2 TTC-BASED EVALUATION

Time to collision (TTC) is defined as the time for two vehicles to collide if two vehicles keep stay in their current path and velocity when the ego vehicle keeps the current lane, which is adopted to assess the driving risk related to the front vehicle collision. The relative distance between the ego vehicle and an obstacle vehicle is decreased with TTC value, and the executing lane-changing distance for the human driver is also decreased, and the driving risk rises. The expression of TTC is defined as

$$ttc = \begin{cases} \dfrac{x_k^e}{v_k^h - v_k^o}, & \dfrac{x_k^e}{v_k^h - v_k^o} > 0 \\ +\infty, & \text{otherwise,} \end{cases} \tag{5.28}$$

where x_k^e is the relative distance from the nearest front obstacle to the ego vehicle, and v_k^h and v_k^o are the longitudinal velocity of the ego vehicle and obstacle vehicle, respectively.

5.3 GAME-BASED COOPERATIVE DRIVING STRATEGY

In this section, a game-based dynamic authority allocation strategy based on the driving safety field is introduced to address path conflicts between the human driver and the path-following controller.

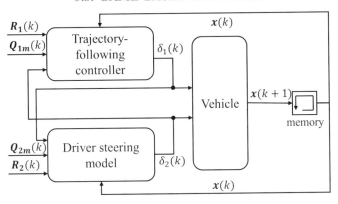

Figure 5.9: Illustration of the game-based cooperative control.

The path conflict is a critical problem in the human-automation cooperative driving system, and the game theory is wildly used to address path conflicts by modeling the human-automation interaction since the game theory has been widely used to address conflicts between two or more players [101]. The framework of the game-based cooperative control is illustrated in Figure 5.9. The non-cooperative Nash game theory is used for the human-automation interaction model. In this framework, the trajectory-following controller compensates the human driver's control signal to decrease possibilities of potential accidents caused by the human driver's wrong actions, and the human driver should estimate the control signal of the controller if he/she still wants to control the vehicle by himself/herself. A dynamic interaction model is built with consideration of the proposed cooperative driving, where controller will control the vehicle to decrease driver's workload under safe condition, however, the driver with the normal driving status still needs to control the vehicle under dangerous conditions in the cooperative control system. Generally, a driver and the trajectory-following controller compensates for each other's control input signal. As a result, the two participants' control strategies converge to the Nash equilibrium of the noncooperative game theory. The non-cooperative Nash strategy and the dynamic authority allocation strategy are presented to describe game-based strategy.

5.3.1   NON-COOPERATIVE NASH EQUILIBRIUM

The non-cooperative Nash equilibrium of the human-automation interaction is derived by the distributed model predictive control (DMPC) method. The 2DOF single-track bicycle vehicle model is adopted for the development of DMPC framework. The 2DOF vehicle model considers the lateral position y and the yaw angle φ, and the vehicle motion are expressed as

$$\begin{cases} m\dot{v} = F_{yf} + F_{yr} - mur \\ I_z \dot{r} = a F_{yf} - b F_{yr}, \end{cases} \tag{5.29}$$

where F_{yf} and F_{yr} represent the lateral tire force of the front and real wheels, respectively, I_z refers to the moment of inertia corresponding to Z-axis of the vehicle, a and b denote the length from the front and rear wheel to the vehicle's CoG, respectively.

Assume that the slip angle of tire is under the small angle, and the tire force is proportional to the small slip angle. Then, the lateral tire force F_{yf} and F_{yr} are calculated by the linear tire model,

$$\begin{cases} F_{yf} = k_f \alpha_f \\ F_{yr} = k_r \alpha_r \end{cases} \tag{5.30}$$

with K_f and K_r are the cornering stiffness of the front and rear tires, respectively. α_f and α_r denote the tire slip angle of the front and real wheels, and calculated by

$$\begin{cases} \alpha_f = \beta + \frac{a\varphi}{u} - \delta \\ \alpha_r = \beta - \frac{b\varphi}{u}, \end{cases} \tag{5.31}$$

where $\beta = v/u$ is the sideslip angle of CoG and δ refers to the steering angle.

For human-automation cooperative driving systems, there provides possibilities for human drivers and automated driving systems to control the vehicle together. Based on the 2DOF vehicle model, the vehicle dynamics of the human-automation system are expressed as

$$\begin{aligned} x(k+1) &= \mathbf{A}x(k) + \mathbf{B}_c \delta_c(k) + \mathbf{B}_d \delta_d(k) \\ z(k) &= \mathbf{C}x(k), \end{aligned} \tag{5.32}$$

where $x(k) = [v(k), \phi(k), X, Y]^T$ are the state variable, $\mathbf{A}$ is the system state matrix and $\mathbf{B}_c$ and $\mathbf{B}_d$ are the input matrices of the trajectory-following controller and the driver steering model. $z(k)$ means the output matrix consisting of the lateral position $y(k)$ and yaw angle $\varphi(k)$. Referring to Equation (4.66), the prediction output of the human-automation system at next N_p time step could be defined as

$$Z(k) = \Phi x(k) + \Theta_c \delta_c(k) + \Theta_d \delta_d(k) \tag{5.33}$$

with

$$\Theta_d = \begin{bmatrix} \mathbf{CB_d} & \mathbf{0} & \cdots & \mathbf{0} \\ \mathbf{CAB_d} & \mathbf{CB_d} & \cdots & \mathbf{0} \\ \vdots & \vdots & \ddots & \vdots \\ \mathbf{CA}^{N_p-1}\mathbf{B_d} & \mathbf{CA}^{N_p-2}\mathbf{B_d} & \cdots & \mathbf{CA}^{N_p-N_u}\mathbf{B_d} \end{bmatrix}.$$

In the framework of the proposed human-automation cooperative driving system, the cost function of each participant is to track their own desired trajectory as close as possible. Thus, the cost function of the trajectory-following controller and the driver are defined as

$$J_c(k) = \|Z(k) - R_c(k)\|_{Q_{1m}}^2 + \|\delta_c(k)\|^2 \tag{5.34}$$

and

$$J_d(k) = \| Z(k) - R_d(k) \|^2_{Q_{2m}} + \| \delta_d(k) \|^2, \tag{5.35}$$

where Q_{1m} and Q_{2m} are the dynamic weight matrices of the trajectory-following controller and the driver, respectively, and they are determined by the dynamic authority allocation strategy discussed in the following part. As illustrated in Figure 5.9, the trajectory-following controller and the drive take the other's control signal into their optimal process, and the two participants converge to the Nash equilibrium which is defined that none of participants can benefit more by unilaterally changing its own control actions [102]. The Nash equilibrium is derived by solving the coupled optimal problem

$$\min_{\delta_c^*} J_c(k) \text{ and } \min_{\delta_c^*} J_c(k) \tag{5.36a}$$

subject to

$$Z(k) = \Phi x(k) + \Theta_c \delta_c(k) + \Theta_d \delta_d(k), \tag{5.36b}$$

where (δ_c^*, δ_d^*) is the Nash solution of two participants in the human-automation cooperative driving system. In order to simplify the coupled optimal problem, the zero-input error ϵ_i is introduced:

$$\begin{aligned} \epsilon_c(k) &= R_c(k) - \Phi x(k) - \Theta_2 \delta_d(k) \\ \epsilon_d(k) &= R_d(k) - \Phi x(k) - \Theta_1 \delta_c(k). \end{aligned} \tag{5.37}$$

By substituting Equation (5.37) to the cost functions (5.34) and (5.35), cost functions can be eliminated as the least-square form

$$J_i(k) = \left\| \begin{bmatrix} S_i \{ \Theta_i \delta_i(k) - \varepsilon_i(k) \} \\ \delta_i(k) \end{bmatrix} \right\|^2, \tag{5.38}$$

where $S_i^T S_i = Q_{im}(k)$ and $i \in \{c, d\}$. The above optimal problem with the least-square form can be solved by the QR algorithm , and the optimal Nash solution of the trajectory-following controller and the driver is calculated by

$$\begin{bmatrix} \delta_c^* \\ \delta_d^* \end{bmatrix} = \begin{bmatrix} K_c^f & 0 \\ 0 & K_d^f \end{bmatrix} \begin{bmatrix} \varepsilon_c(k) \\ \varepsilon_d(k) \end{bmatrix} \tag{5.39}$$

with

$$K_i^f(k) = \begin{bmatrix} S_i \Theta_i \\ I \end{bmatrix} \setminus \begin{bmatrix} S_i \\ 0 \end{bmatrix},$$

where I is an identity matrix. Obviously, the control sequences of the participant has influences on the optimal control signal of the other participant in the human-automation cooperative driving system. The convex iteration method is adopted to solve this problem, then the Nash solution of the non-cooperative Nash solution is calculated by

$$\begin{bmatrix} \delta_1^*(k) \\ \delta_2^*(k) \end{bmatrix} = \boldsymbol{K}^{Nash}(k) \begin{bmatrix} x_1(k) \\ x_2(k) \end{bmatrix} \tag{5.40}$$

with

$$x_i(k) = \begin{bmatrix} x(k) R_i(k) \end{bmatrix}^T, \ \Gamma_i = \begin{bmatrix} -\boldsymbol{K}_i^f(k) \boldsymbol{\Phi} \ \boldsymbol{K}_i^f(k) \end{bmatrix}, \ \Lambda_i = \boldsymbol{K}_i^f(k) \boldsymbol{\Theta}_j$$

$$\boldsymbol{K}^{Nash}(k) = \begin{bmatrix} (\mathbf{I} - \Lambda_1 \Lambda_2)^{-1} \Gamma_1 & (\mathbf{I} - \Lambda_1 \Lambda_2)^{-1} \Lambda_1 \Gamma_2 \\ (\mathbf{I} - \Lambda_2 \Lambda_1)^{-1} \Lambda_2 \Gamma_1 & (\mathbf{I} - \Lambda_2 \Lambda_1)^{-1} \Gamma_2 \end{bmatrix}.$$

5.3.2   DYNAMIC AUTHORITY ALLOCATION STRATEGY

In the dynamic authority allocation strategy, there is an assumption that the driver is under the normal driving states. And the game-based strategy is designed to model the interaction that the controller controls the vehicle to decrease driver's workload under safe driving conditions, however, the driver needs to control the vehicle under dangerous driving conditions. Therefore, the objective of the proposed strategy is to adjust the control authority of the controller according to the driving risk conditions. As illustrated in the cost function of Equation (5.34), the interest of the control authority can be indicated by the penalization of the deviation to the desired target path. The higher values of the path-error weight matrices $Q_{1m}(k)$ or $Q_{2m}(k)$ at each time step are, the larger effort the human driver or the controller would pay to follow their desired path. Therefore, the dynamic authority allocation could be realized by adjusting the weight value of the path-error weight matrices in the design of the controller objective function in real time, and it can be defined as

$$\boldsymbol{Q}_{2m}(k) = \lambda(|\boldsymbol{D}_R(k)|) \boldsymbol{Q}_{1m}(k), \tag{5.41}$$

where $\boldsymbol{D}_R(k)$ is the driving risk evaluated by Equation (5.27), and λ is the authority allocation ratio. The basic principles of $\lambda(\cdot)$ are concluded that the more dangerous the driving risk grows, the larger the authority allocation ratio should be and the higher control authority the human driver occupies. As for the authority allocation ratio, it is a single input-single output (SI-SO) system, the look-up table method is adopted. A reference of the look-up table method is listed in Table 5.3.

5.3.3   CASE STUDY

To validate the proposed game-based cooperative control with dynamic authority allocation strategy, the case study provides the comparison results among the proposed strategy, the constant strategy that the human driver and the trajectory-following controller have the scaled control authority, and driver alone without control strategy. As shown in Figure 5.10, there are two road segments, including the obstacle area and obstacle-free area (simulation time $t > 40$ s). The ego vehicle keeps running at 20 m/s and its initial position is at the middle point of the

Table 5.3: Reference look-up table of the authority allocation ratio

$|D_R|$	-400	-350	-300	-250	-200	-150	-100	-50	0
$\ln(\lambda)$	-4.27	-4.27	-4.27	-4.25	-3.56	-2.82	-2.35	-2.01	-1.78
$|D_R|$	50	100	150	200	250	300	350	400	450
$\ln(\lambda)$	-1.42	-0.83	-0.42	0.00	0.21	0.63	1.17	1.61	1.83
$|D_R|$	500	550	600	650	700	750	800	850	900
$\ln(\lambda)$	2.21	2.34	2.56	2.82	3.34	3.83	4.23	4.32	4.67
$|D_R|$	950	1,000							
$\ln(\lambda)$	4.67	4.67							

right-hand lane. As illustrated in Figure 5.10a, three obstacles with different initial states run in different lanes. SO is a static obstacle, which is fixed at the position (140,-2), DO_1 is a dynamic obstacle vehicle initiated at (220,2) and runs along the middle line of the left-hand lane at constant speed at 5 m/s, and DO_2 moves along the middle line of the right-hand lane from the position (260,-2) at the constant velocity of 10 m/s. The vehicle path under the proposed game-based shared control is smooth and the proposed shared control system can avoid obstacles safely. Figure 5.10b,c show details of vehicle paths and driving risk distribution from $t = 0$ s to $t = 9$ s and from $t = 16$ s to $t = 24$ s. As observed, the driving risk is higher in the condition of Figure 5.10c than the condition of (b), since two dynamic obstacle DO_1 and DO_2 have influence on the ego vehicle. Therefore, the human driver occupies higher control authority than the controller under the latter condition, and it also illustrated in Figure 5.10d. The authority allocation ration value in the marked rectangle (c) is larger than the rectangle (b).

After the ego vehicle runs over 40 s, the vehicle comes into the obstacle-free area. According to the dynamic authority allocation strategy, the vehicle is controlled by the trajectory-following controller to decrease driver's workload. As shown in Figure 5.10e, the steering wheel angle of the proposed shared control strategy is smaller than the driver alone and the scaled constant strategy in the obstacle-free area that is marked as (I). Besides, the vehicle under the proposed strategy can obtain better stability performance than other two strategy since its lateral acceleration value is smaller, as depicted in Figure 5.10f.

5.4   FUZZY-BASED COOPERATIVE DRIVING STRATEGY

Game-based shared control strategy is applied into the condition that the human driver is under the normal driving status. However, it's also critical to deal with the potential collision situation caused by the human driver's fatigue or distracted status. Therefore, the shared control strategy will be designed by using the fuzzy control method considering the driver's driving intention,

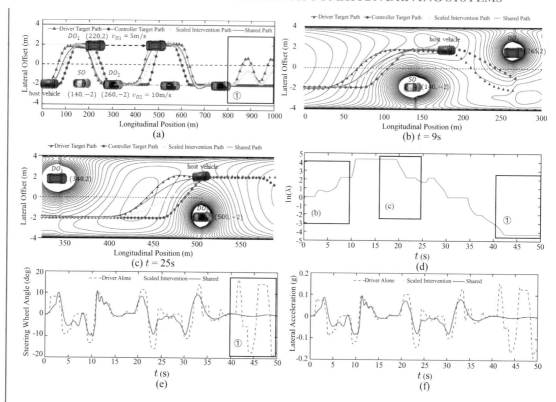

Figure 5.10: Results of the game-based strategy: (a) illustration of the case scenario; (b) presentation of the elliptic driving safety field distribution and vehicle paths at $t = 9$ s; (c) presentation of the elliptic driving safety field distribution and vehicle paths at $t = 25$ s; (d) authority allocation ratio; (e) steering wheel angle; and (f) lateral acceleration.

driving risk and performance index. The evaluated driving risk, performance index, and the fuzzy-based strategy will be introduced.

5.4.1   FRAMEWORK ILLUSTRATION

The proposed framework of the fuzzy-based cooperative driving strategy is illustrated in Figure 5.11. The information of environment and vehicle state are obtained by sensors, such as camera, radar, and GPS [103]. The obstacle avoidance path is planned by the artificial potential field methods presented in Section 3.3 or the optimal path-planning with natural cubic splines in Section 3.4. To recognize human drivers' driving intention, drivers' steering control signal first will be input into an 8DOF vehicle dynamics model to generate the vehicle path driven by the driver alone. Then, the steering wheel angle and vehicle lateral position of driver alone are used for driving intention recognition by using machine learning method, such as the SVM

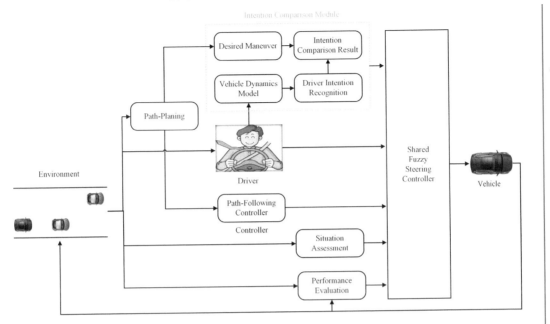

Figure 5.11: Illustration of the fuzzy-based cooperative driving strategy.

method (Section 5.1.2) or the iMLCU approach (Section 5.1.3). After obtaining the driver's driving intention, it is used for the comparison with desired maneuvers, and the comparison result is employed for the development of the fuzzy-based cooperative driving strategy. The situation assessment and performance index are also used for the strategy development, and the cooperative coefficient denoting the control authority of human drivers and the controller and the final shared steering command are determined.

5.4.2 PERFORMANCE INDEX

The other parameter for the fuzzy-based shared control strategy is the performance index that is adopted to evaluate the path-following performance of the shared control system and the human driver. The larger the lateral distance to the desired path is, the worse the performance index grows. Obviously, the larger lateral deviation caused by a human driver driving alone is, the higher proportion assistance the human driver needs, and the bigger lateral deviation of the shared system is, the higher control proportion the path-following controller occupies to correct the path error. The performance index is defined as

$$\eta = \begin{cases} \dfrac{\varepsilon}{y_l}, & \varepsilon \le y_l \\ 1, & \text{otherwise} \end{cases} \tag{5.42}$$

with

$$\varepsilon = \frac{|\Delta y + \lambda \Delta y_p|}{2},$$

where y_l is a upper threshold of the lateral deviation, Δy_p means the predicted lateral deviation between the path of the human driver observed by inputted into the 8DOF vehicle model and the desired path, and λ refers to the compensation constant for the prediction error.

5.4.3   DRIVER'S DRIVING INTENTION AND DESIRED MANEUVER

As the ego vehicle is controlled by the cooperative driving system, the final shared control signal inputted to the ego vehicle is the blend of driver's and the controller's. Thus, both the human driver and the controller have influences on the vehicle control. The recognized intention based on y_{rc} and $v_{y_{rc}}$ of the host vehicle measured by sensors cannot reflect real driver's driving intention. Therefore, in order to recognize the driver's lane-changing intention, the accurate prediction for vehicle global lateral position driven by a human driver alone is necessary. Driver's steering wheel angle can be obtained by sensors, thus, an accurate vehicle model is needed to predict the vehicle lateral position. An 8DOF four-wheel vehicle model could provide a great accuracy level [104], and the driver's driving intention will be recognized by steering wheel angle and predicted vehicle position by using machine learning method.

The 8DOF vehicle model is introduced briefly here. The 8DOF vehicle model includes the longitudinal position x, lateral position y, yaw angle φ, roll angle ϕ, and the four-wheel rotational speeds $(\omega_{fl}, \omega_{fl}, \omega_{fr}, \omega_{rl}, \omega_{rr})$. As shown in Figure 5.12, the following equations are used to describe the motion of the 8DOF vehicle model

$$\begin{cases} m(\dot{u} - v\dot{\varphi}) - m_s h \phi \dot{\varphi} = F_{xfl} + F_{xfr} + F_{xrl} + F_{xrr} \\ m(\dot{v} + u\dot{\varphi}) + m_s h \ddot{\phi} = F_{yfl} + F_{yfr} + F_{yrl} + F_{yrr} \\ I_x \ddot{\phi} + m_s h(\dot{v} + u\dot{\varphi}) = M_x \\ I_z \ddot{\varphi} = M_z \\ \dot{\omega}_{fl} = \frac{1}{J_{wL}}(-F_{xfl} r_{eff} - T_{brk} - M_{fr1}) \\ \dot{\omega}_{fr} = \frac{1}{J_{wL}}(-F_{xfr} r_{eff} - T_{brk} - M_{fr2}) \\ \dot{\omega}_{rl} = \frac{1}{J_{wR}}(-F_{xrl} r_{eff} - T_{brk} - M_{fr3}) \\ \dot{\omega}_{rr} = \frac{1}{J_{wR}}(-F_{xrr} r_{eff} - T_{brk} - M_{fr4}), \end{cases} \tag{5.43}$$

where m is the lumped mass, m_s refers to the sprung mass. I_x means the moment of inertia corresponding to the vehicle Z and x axis. J_{wi} represent the equivalent inertia for each tire. r_{eff} refers to the effective radius of the tire. T_{brk} is the braking torque acting on each tire. M_{fri} is the rolling resistance torque. F_{xi} and F_{yi}, $i \in \{fl, fr, rl, rr\}$, represent the longitudinal and lateral

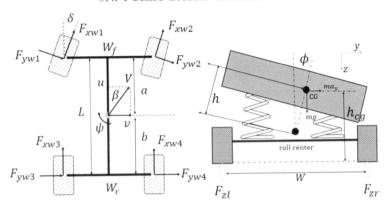

Figure 5.12: Illustration of an 8DOF vehicle model.

force of each wheel, respectively, and denoted by

$$
\begin{bmatrix} F_{xi} \\ F_{yi} \\ F_{xj} \\ F_{yj} \end{bmatrix} = \begin{bmatrix} \cos\delta & -\sin\delta & 0 & 0 \\ \sin\delta & \cos\delta & 0 & 0 \\ 0 & 0 & 1 & 0 \\ 0 & 0 & 0 & 1 \end{bmatrix} \begin{bmatrix} F_{xti} \\ F_{yti} \\ F_{xtj} \\ F_{ytj} \end{bmatrix}, \; i \in \{fl, fr\}, \; j \in \{rl, rr\},
\tag{5.44}
$$

where F_{xt} and F_{yt} are the tire forces in the X and Y directions, respectively, which are calculated by the Magic formula tire model (Section 4.3.1) described in the above part. Moreover, M_x and M_z are the moment respecting to X and Z axis, and they are calculated by

$$
M_x = -2(D_{\phi f} + D_{\phi r})\dot\phi + (m_s g h - 2(K_{\phi f} + K_{\phi r}))\phi,
\tag{5.45}
$$

$$
M_z = \frac{1}{2}\left(-W_f(F_{xfl} + F_{xrl}) + W_r(F_{xfr} + F_{xrr})\right) + a(F_{yfl} + F_{yfr}) - b(F_{yrl} + F_{yrr}),
\tag{5.46}
$$

where W_f and W_r refer to the front and rear wheelbase, respectively, $K_{\phi f}$ and $K_{\phi r}$ represent the front and rear chassis stiffness, and $D_{\phi f}$ and $D_{\phi r}$ denote the front and rear chassis rolling stiffness.

The desired lane-changing maneuver is determined according to the desired path for obstacle avoidance and the road centerline. The relative lateral position to the road centerline is denoted as

$$
y_{os}(k) = \mathbf{P}(\mathbf{y_r}(k) - \mathbf{r_c}(k))
\tag{5.47}
$$

with

$$
\mathbf{y_r}(k) = \begin{bmatrix} y_{ref}(k) \\ y_{ref}(k+1) \\ \vdots \\ y_{ref}(k+N_p) \end{bmatrix} \quad \mathbf{r_c}(k) = \begin{bmatrix} c r_c(k) \\ r_c(k+1) \\ \vdots \\ r_c(k+N_p) \end{bmatrix},
$$

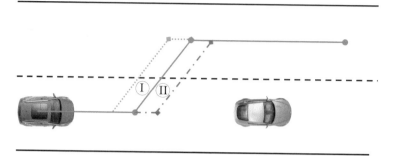

Figure 5.13: Illustration of two inconsistent conditions.

where $\mathbf{P} = \begin{bmatrix} 1 & 0 & \cdots & 0 \end{bmatrix}$ is the matrix to select the first element. Thus, the relative lateral velocity to the road center line is expressed as follows:

$$v_{y_{os}}(k) = \frac{1}{T_s}(y_{os}(k) - y_{os}(k-1)). \tag{5.48}$$

Then, the sample data $[y_{os}(k) \ v_{y_{os}}(k)]^T$ are inputed into the iMLCU classifier to predict the desired maneuver.

5.4.4   FUZZY-BASED STRATEGY

The final steering command u_s of the fuzzy-based strategy is a blend of the human driver's input u_{dr} and the path-following controller's input u_c, which is defined as

$$u_s = ku_c + (1-k)u_{dr}, \tag{5.49}$$

where k refers to the cooperative coefficient determined by the fuzzy-based strategy.

The purpose of the fuzzy-based shared control strategy is to determine the authority proportion of the path-following controller according to the driving safety level, performance index, and the comparison result between the driver's lane-changing intention and the desired maneuvers. When the driver's lane-changing intention is compared with the desired maneuver, there are three conditions, including the consistent condition, and the other two inconsistent conditions, as shown in Figure 5.13. The first inconsistent compared condition is the advanced lane-changing intention where the human driver tends to change the before the desired lane-changing longitudinal point x_{de}, which is marked as (I). And the first inconsistent condition makes the driver safer to avoid obstacle by lane-changing, and the controller's control authority should be weakened when compared with the consistent condition. In the same way, the second inconsistent condition marked as (II) is the lagged condition where the human driver tends to change the lane behind the desired point x_{de}. In this condition, the controller's control authority should be strengthened since it's a more dangerous maneuver than the consistent condition.

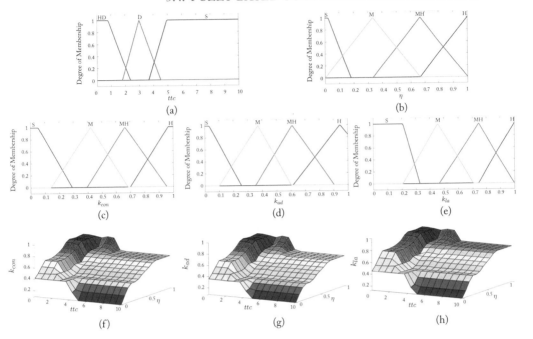

Figure 5.14: The illustration of the shared fuzzy controller: (a) membership function of *ttc*; (b) membership function of the performance evaluation index; (c) membership function of the cooperative coefficient in the consistent intention condition k_{con}; (d) membership function of the cooperative coefficient in the advanced inconsistent intention condition k_{ad}; (e) membership function of the cooperative coefficient in the lagged inconsistent intention condition k_{la}; (f) fuzzy surface of the consistent fuzzy controller; (g) fuzzy surface of the advanced inconsistent fuzzy controller; and (h) fuzzy surface of the lagged inconsistent fuzzy controller.

Therefore, there are three fuzzy-based shared controller based on the driving risk and performance index, namely, the consistent, the advanced inconsistent, and the lagged inconsistent fuzzy controller. Three fuzzy controller are designed as follows.

Fuzzy Variables

Fuzzy variables consist of two input variables and three output variables of the fuzzy-based shared controller. The input variables include *ttc* and η calculated from Equations (5.28) and (5.42), respectively. The cooperative coefficients of three fuzzy-based shared controllers are defined as output variables, including k_{con}, k_{ad}, and k_{la}.

ttc has three associated linguistic values, including high dangerous (HD), dangerous (D), and safe (S), which are designed according to driving risk evaluated by *ttc*. The membership function shape is illustrated in Figure 5.14a. η has four associated linguistic values, namely, small

Table 5.4: Shared fuzzy controller rule bases

η / ttc	S	M	MH	H
HD	M	MH	H	H
D	M	M	MH	H
S	S	M	MH	MH

(S), medium (M), medium high (MH), and high (H), and its membership function is shown in Figure 5.14b.

k_{con}, k_{ad}, and k_{la} are the cooperative coefficients of the consistent, advanced inconsistent, and lagged inconsistent fuzzy-based shared controller, respectively, and they have four associated linguistic values, including S, M, MH, and H. To meet the requirement of the fuzzy-based strategy, the value of k_{la} should be bigger than the value of k_{con}, and the value of k_{con} should be larger than the value of k_{ad}. The membership functions of k_{con}, k_{ad}, and k_{la} are shown in Figure 5.14c–e.

Fuzzy Rules and Inference

Fuzzy rules and inference are critical parts of the fuzzy controller. Fuzzy rules are generated based on expert knowledge and it also followed the principle that the cooperative coefficient increases with driving risk and performance index η. All the fuzzy rules are listed in Table 5.4.

Besides, the Mamdani inference approach is used to handle fuzzy implication problem and the inference process is defined as follows [105]:

$$\text{IF } ttc \text{ is } A \text{ and } \eta \text{ is } B \text{ THEN } k \text{ is } C, \tag{5.50}$$

where A, B, and C are the fuzzy linguistic values of input and output variables. The fuzzy surfaces of three fuzzy-based shared controller are shown in Figure 5.14f–h.

Defuzzification

Defuzzification is the process that maps the fuzzy linguistic value to classical value applied into the system control. The center-of-area (CoA) defuzzification method is adopted, which is

$$CoA = \frac{\sum_i \mu(x_i) x_i}{\sum_i \mu(x_i)}, \tag{5.51}$$

where $\mu(x_i)$ is the membership function value for the linguistic variable point x_i.

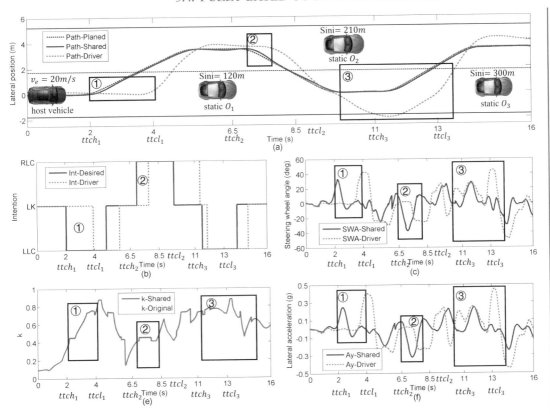

Figure 5.15: Results of the fuzzy-based cooperative driving system: (a) illustration of the case scenario and vehicle paths; (b) driving intention and desired maneuver; (c) steering wheel angle; (d) cooperative coefficient; and (e) lateral acceleration.

5.4.5   CASE STUDY

Figure 5.15 gives case study of the proposed fuzzy-based cooperative driving strategy. To validate the proposed strategy, there are two different driving status provided, including the normal and distracted driving status. As shown in Figure 5.10a, there are three obstacles on a straight road, and the human driver will receive a phone call after the vehicle exceeds over the second obstacle vehicle, which makes the driver in a distracted driving status to avoid the third obstacle. The host vehicle keeps moving with a constant speed at 20 m/s. The first O_1 and third obstacle O_3 are fixed at the middle point of the right-hand lanes, and the initial position of the first and third obstacle are 120 m and 300 m far from the original position. The second obstacle O_2 is fixed at the middle of the left-hand lane with 210 m far from the original position. Besides, there are also provided vehicle paths driven by a naive young driver and the fuzzy-based cooperative driving system. The mean lateral position error to the planned path of the driver is 0.73 m with 0.59

RMSE value while the mean error of the fuzzy-based strategy is 0.04 m with 0.02 RMSE value. Therefore, the fuzzy-based strategy can realize better path-following performance than a driver. Driver's driving intention and desired maneuver are shown in Figure 5.15b, and there are several inconsistent area during the case study. The cooperative coefficient can be adjusted higher or lower according to fuzzy-based strategy, as illustrated in Figure 5.15e. After exceeding over O_2, a driver received a phone call, which made him distracted. The fuzzy-based strategy can help the distracted driver avoid obstacle O_3 safely, which can be proved by the path shown in Figure 5.15a. In addition, the fuzzy-based strategy help the vehicle improve its stability performance since the maximum value of the lateral acceleration is 0.2471 g and it is lower than the value of the human-driven vehicle.

5.5   SUMMARY

In this chapter, human-automation cooperative driving systems are introduced. Driving intention and situation assessment are two main factors that are considered into the design of human-automation cooperative driving systems. The process of the data collecting experiment for driving intention recognition is described, and two machine learning methods, SVM and iMLCU approaches, are used to train the driver's lane-changing intention recognition model. Driving safety field and TTC-based evaluation methods are adopted for the driving risk assessment by considering driver-vehicle-road interactions. Two cooperative driving strategies are proposed to address human-automation conflicts. Game-based strategy is designed based on the game theory and a dynamic authority allocation strategy to deal with human-automation path conflicts. Fuzzy-based strategy is proposed based on the driving intention and fuzzy control theory to solve authority conflicts. Both strategies can help human drivers safely avoid obstacles.

Bibliography

[1] World Health Organization (WHO). Violence injury prevention. https://www.who.int/health-topics/road-safety#tab=tab_1 1

[2] World Health Organization (WHO). Improving global road safety. http://www.who.int/roadsafety-/about/resolutions/download/en/index.html 1

[3] National Bureau of Statistics of China, 24-5 Basic Statistics on Traffic Accidents (2018), China Statistical Yearbook, 2019. http://www.stats.gov.cn/tjsj/ndsj/2019/indexeh.htm 1

[4] National Center for Statistics and Analysis. Traffic safety facts 2015 data passenger vehicle. Washington, DC, National Highway Traffic Safety Administration, 2017, Report no. DOT HS 812 413. 1

[5] Bellman, R. *Dynamic Programming*, Princeton University Press, New York, 1957. DOI: 10.1126/science.153.3731.34. 2

[6] Werling, M., Kammel, S., Ziegler, and J., Gröll, L. Optimal trajectories for time-critical street scenarios using discretized terminal manifolds. *Int. J. Robot. Res.*, 31(3):346–359, 2012. DOI: 10.1177/0278364911423042. 2

[7] Gerdts, M., Karrenberg, S., Müller-Bessler, B., and Stock, G. Generating locally optimal trajectories for an automatically driven car. *Optim. Eng.*, 10(4):439–463, 2009. DOI: 10.1007/s11081-008-9047-1. 2

[8] McNaughton, M. Parallel algorithms for real-time motion planning. Ph.D. Thesis, Carnegie Mellon University, 2011. 2

[9] Gu, T. and Dolan, J. M. On-road motion planning for autonomous vehicles. *International Conference on Intelligent Robotics and Applications*, Springer, Berlin, Heidelberg, 2012. DOI: 10.1007/978-3-642-33503-7_57. 2

[10] Buehler, M., Iagnemma, K., and Singh, S. *The DARPA Urban Challenge: Autonomous Vehicles in City Traffic*, Berlin, Springer, 2009. DOI: 10.1007/978-3-642-03991-1. 3

[11] Li, X., Sun, Z., Cao, D., et al. Real-time trajectory planning for autonomous urban driving: Framework, algorithms, and verifications. *IEEE/ASME Transactions on Mechtronics*, 21(2):740–753, 2015. DOI: 10.1109/tmech.2015.2493980. 3

[12] Naranjo, J. E., Gonzalez, C., and Garcia, R. Lane-change fuzzy control in autonomous vehicles for the overtaking maneuver. *IEEE Transactions on Intelligent Transportation Systems*, 9(3):438–450, 2008. DOI: 10.1109/tits.2008.922880. 3

[13] Kodagoda, K. R. S., Wijesoma, W. S., and Teoh, E. K. Fuzzy speed and steering control of an AGV. *IEEE Transactions on Control Systems Technology*, 10(1):112–120, 2002. DOI: 10.1109/87.974344. 3

[14] Moon, S., Moon, I., and Yi, K. Design, tuning, and evaluation of a full-range adaptive cruise control system with collision avoidance. *Control Engineering Practice*, 17(4):442–455, 2009. DOI: 10.1016/j.conengprac.2008.09.006. 3

[15] Bu, F., Tan, H. S., and Huang, J. J. Design and field testing of a cooperative adaptive cruise control system. *Proc. of the American Control Conference*, pages 4616–4621, IEEE, 2010. DOI: 10.1109/acc.2010.5531155. 3

[16] Khatib, O. Real-time ostacle avoidance for manipulators and mobile robots. *International Journal of Robotics Research*, 5(1):90–98, 1986. DOI: 10.1109/robot.1985.1087247. 3

[17] Brandt, T., Sattel, T., and Wallaschek, J. Towards vehicle trajectory planning for collision avoidance based on elastic bands. *International Journal of Vehicle Autonomous Systems*, 5(1–2):28–46, 2007. DOI: 10.1504/ijvas.2007.014928. 4

[18] Sattel, T. and Brandt, T. From robotics to automotive: Lane-keeping and collision avoidance based on elastic bands. *Vehicle Systems Dynamics*, 46(7):597–619, 2008. DOI: 10.1080/00423110701543452. 4

[19] Cao, H., Song, X., Huang, Z., and Pan, L. Simulation research on emergency path planning of an active collision-avoidance system combined with longitudinal control for an autonomous vehicle. *Proc. IMechE D, Journal of Automobile Engineering*, 230(12):1624–1653, 2016. DOI: 10.1177/0954407015618533. 4

[20] Mingjun, L., Song, X., Cao, H., and Huang, Z. Shared steering control combined with driving intention for vehicle obstacle avoidance. *Proc. of the Institution of Mechanical Engineers, Proc. IMechE D, Journal of Automobile Engineering*, 233(11):2791–2808. DOI: 10.1177/0954407018806147. 4, 13

[21] Ji, J., Khajepour, A., Melek, W. W., and Huang, Y. Path planning and tracking for vehicle collision avoidance based on model predictive control with multi-constraints. *IEEE Transactions on Vehicle Technology*, 66(2):952–964, 2016. DOI: 10.1109/tvt.2016.2555853. 4

[22] Rasekhipour, Y., Khajepour, A., Chen, S.-K., and Litkouhi, B. A potential field-based model predictive path-planning controller for autonomous road vehicles.

IEEE Transactions on Intelligent Transportation Systems, 18(5):1255–1267, 2016. DOI: 10.1109/tits.2016.2604240. 4

[23] Lavalle, S. M. and Kuffner, J. J. Randomized kinodynamic motion planning. *International Journal of Robotics Research*, 20(5):378–400, 2001. 4

[24] Lavalle, S. M. From dynamic programming to RRTs: Algorithmic design of feasible trajectories. *Springer Tracts in Advanced Robotics*, 4:19–37, 2003. DOI: 10.1007/3-540-36224-x_2. 4

[25] Lavalle, S. Rapidly exploring random trees: Progress and prospects. *4th Workshop on the Algorithmic Foundations of Robotics; Algorithmic and Computational Robotics*, pages 293–308, Hanover, 2000.

[26] Lai-Chun, F., Hua-Wei, L., Ming-Bo, D., et al. Guiding-area RRT path planning algorithm based on A* for intelligent vehicle. *Computer Systems Applications*, 26(8):127–133, 2017. 4

[27] Ma, L., Xue, J., Kawabata, K., et al. A fast RRT algorithm for motion planning of autonomous road vehicles. *17th International IEEE Conference on Intelligent Transportation Systems*, IEEE, pages 1033–1038, 2014. DOI: 10.1109/itsc.2014.6957824. 4

[28] Song, X., Zhou, N., Huang, Z., et al. An improved RRT algorithm of local path planning for vehicle collision avoidance. *Journal of Human University (Natural Sciences)*, 44(4):30–37, 2017. 5

[29] Yuan, C., Liu, G., and Zhang, W. 6-DOF industrial manipulator motion planning based on RRT-connect algorithm. Tan J. (Ed.), *Advances in Mechanical Design, ICMD. Mechanisms and Machine Science*, 77, Springer, Singapore, 2020. DOI: 10.1007/978-981-32-9941-2_8. 5

[30] Colorni, A., Dorigo, M., and Maniezzo, V. Distributed optimization by ant colonies. Toward a practice of autonomous systems. *Proc. of the 1st European Conference on Artificial Life*, Cambridge, MIT Press, 1992. 5

[31] Fan, X., Luo, X., Yi, S., et al. Path planning for robots based on ant colony optimization algorithm under complex environments. *Control and Decision*, 19(2):166–170, 2004. 5

[32] Wu, T., Xu, J., Liu, J., et al. Multi-strategy ant colony algorithm for cross-country path planning. *Journal of PLA University of Science and Technology (Natural Science Edition)*, 15(2):158–164, 2014. 5

[33] Liang, X., Wang, H., Meng, G., et al. Path planning for UAV under three-dimensional real terrain in the rescue mission. *Journal of Beijing University of Aeronautics and Astronautics*, 41(7):1183–1187, 2015. 5

[34] Jabbarpour, M. R., Zarrabi, H., Jung, J. J., et al. A green ant-based method for path planning of unmanned ground vehicles. *IEEE Access*, (5):1820–1832, 2017. DOI: 10.1109/access.2017.2656999. 5

[35] Lee, K.-T., et al. Realization of an energy-based ant colony optimization algorithm for path planning. *New Trends on System Science and Engineering: Proceedings of ICSSE*, pages 193–199, Japan, 2015. 6

[36] Cong, Y. Z. and Ponnambalam, S. G. Mobile robot path planning using ant colony optimization. *IEEE/ASME International Conference on Advanced Intelligent Mechatronics*, pages 851–856, 2009. DOI: 10.1109/aim.2009.5229903. 6

[37] Li, X. A new type of intelligent optimization method-artificial fish scholas algorithm. Ph.D. Thesis, Zhejiang University, Hangzhou, 2003. 6

[38] Yao, Z., Ren, Z., and Chen, Y. Path planning for mine rescue robot based on AFSA. *Coal Mine Machinery*, 35(4):59–61, 2014. 6

[39] Ma, W., Wu, Z., Yang, J., et al. Decision support from artificial fish swarm algorithm for ship collision avoidance route planning. *Navigation of China*, 37(3):63–67, 2014. 6

[40] Zhang, W., Lin, Z., Liu, T., et al. Robot path planning method based on modified artificial fish swarm algorithm. *Computer Simulation*, 33(12):374–379, 2016. DOI: 10.1155/2016/3297585. 6

[41] Shah-Hosseini, H. Problem solving by intelligent water drops. *IEEE Congress on Evolutionary Computation*, pages 3226–3231, 2007. DOI: 10.1109/cec.2007.4424885. 6

[42] Shah-Hosseini, H. Intelligent water drops algorithm: A new optimization method for solving the multiple knapsack problem. *International Journal of Intelligent Computing and Cybernetics*, 1(2):193–212, 2008. DOI: 10.1108/17563780810874717. 7

[43] Kamkar, I., Akbarzadeh-T. M. R., and Yaghoobi, M. Intelligent water drops a new optimization algorithm for solving the vehicle routing problem. *Systems Man and Cybernetics (SMC), IEEE International Conference on*, pages 4142–4146, 2010. DOI: 10.1109/icsmc.2010.5642405. 7

[44] Duan, H., Liu, S., and Lei, X. Air robot path planning based on intelligent water drops optimization. *International Joint Conference on Neural Networks*, pages 1397–1401, 2008. DOI: 10.1109/ijcnn.2008.4633980. 7

[45] Duan, H., Liu, S., and Wu, J. Novel intelligent water drops optimization approach to single UCAV smooth trajectory planning. *Aerospace Science and Technology*, 13(8):442–449, 2009. DOI: 10.1016/j.ast.2009.07.002. 7

[46] Song, X., Pan, L., and Cao, H. Local path planning for vehicle obstacle avoidance based on improved intelligent water drops algorithm. *Automotive Engineering*, 38(2):185–191, 228, 2016. 7

[47] Mcruer, D. T. and Krendel, E. S. The human operator as a servo system element. *Journal of the Franklin Institute*, 267(6):511–536, 1959. DOI: 10.1016/0016-0032(59)90072-9. 7

[48] Weir, D. H. and Mcruer, D. T. Dynamics of driver-vehicle steering control. *Automatica*, 6(1):87–98, 1970. DOI: 10.1016/0005-1098(70)90077-4. 7

[49] Mcruer, D. T., Allen, R. W., Weir, D. H., et al. New results in driver steering control models. *Human Factors*, 19(4):381–397, 1977. DOI: 10.1177/001872087701900406. 7

[50] McRuer, D. T. and Krendel, E. Mathematical models of human pilot behavior. *Advisory Group on Aerospace Research and Development AGARDograph 188*, January 1974. 7

[51] Macadam, C. C. Application of an optimal preview control for simulation of closed-loop automobile driving. *IEEE Transactions on Systems Man and Cybernetics*, 11(6):393–399, 2007. DOI: 10.1109/tsmc.1981.4308705. 8

[52] Macadam, C. C. An optimal preview control for linear systems. *Journal of Dynamic Systems Measurement and Control*, 102(3):188–190, 1980. DOI: 10.1115/1.3139632. 8

[53] Guo, K. Development of a longitudinal and lateral driver model for autonomous vehicle control. *International Journal of Vehicle Design*, 36(1):50–65, 2004. DOI: 10.1504/ijvd.2004.005320. 9

[54] Guo, K., Cheng, Y., and Ding, H. Analytical method for modeling driver in vehicle directional control. *18th IAVSD Symposium: The Dynamics of Vehicles on Roads and Tracks*, pages 401–410, Kanagawa-ken (Japan), 2003. 9

[55] Ding, H., Guo, K., Li, F., et al. Arbitrary path and speed following driver model based on vehicle acceleration feedback. *Journal of Mechanical Engineering*, 46(10):116–120, 2010. DOI: 10.3901/jme.2010.10.116. 9

[56] Sharp, R. S. and Valtetsiotis, V. Optimal preview car steering control. *Vehicle Systems Dynamics*, 35(1):101–117, 2001. 10

[57] Sharp, R. S. Driver steering control, and a new perspective on car handling qualities. *Proc. IMechE C, Journal of Mechanical Engineering Science*, 219(219):1041–1051, 2005. DOI: 10.1243/095440605x31896. 10

[58] Timings, J. P. and Cole, D. J. Vehicle trajectory linearization to enable efficient optimization of the constant speed racing line. *Vehicle Systems Dynamics*, 50(6):883–901, 2012. DOI: 10.1080/00423114.2012.671946. 12

[59] Timings, J. P. and Cole, D. J. Minimum maneuver time calculation using convex optimization. *Journal of Dynamic Systems, Measurement, and Control*, 135, 031015-1:9, 2013. DOI: 10.1115/1.4023400. 12

[60] Raffo, G. V., Gomes, G. K., and Normey-Rico, J. E. A predictive controller for autonomous vehicle path tracking, *IEEE Transactions on Intelligent Transportation Systems*, 10(1):92–102, 2009. DOI: 10.1109/tits.2008.2011697. 12

[61] Falcone, P., Borrelli, F., Asgari, J., et al. Predictive active steering control for autonomous vehicle systems, *IEEE Transactions on Control Systems Technology*, 15(3):566–580, 2007. DOI: 10.1109/tcst.2007.894653. 12

[62] Arikere, A., Yang, D., Klomp, M., et al. Integrated evasive maneuver assist for collision mitigation with oncoming vehicles. *Vehicle Systems Dynamics*, 56(10):1577–1603, 2018. DOI: 10.1080/00423114.2017.1423091. 12

[63] Ghazali, M., Durali, M., and Salarieh, H. Path-following in model predictive rollover prevention using front steering and braking. *Vehicle Systems Dynamics*, 55(1):121–148, 2017. DOI: 10.1080/00423114.2016.1246741. 12

[64] Zeyada, Y., El-Beheiry, E., El-Arabi, M., et al. Driver modeling using fuzzy logic controls for human-in-the-loop vehicle simulations, *Current Advances in Mechanical Design and Production VII*, Edited by M. F. Hassan and S. M. Megahed (Elsevier Science Ltd. Oxford), 2000. DOI: 10.1016/b978-008043711-8/50009-x. 12

[65] Lin, Y., Tang, P., Zhang, W. J., et al. Artificial neural network modeling of driver handling behavior in a driver-vehicle-environment system. *International Journal of Vehicle Design*, 37(1):24–45, 2005. DOI: 10.1504/ijvd.2005.006087. 12

[66] Hu, C., et al. Robust H∞ output-feedback control for path following of autonomous ground vehicles. *Mechanical Systems and Signal Processing*, 70:414–427, 2016. DOI: 10.1109/cdc.2015.7402425. 12

[67] Bifulco, G. N., Pariota, L., Simonelli, F., et al. Development and testing of a fully adaptive cruise control system. *Transportation Research Part C: Emerging Technologies*, 29:156–170, 2013. DOI: 10.1016/j.trc.2011.07.001. 12

[68] Bifulco, G. N., Pariota, L., Brackstione, M., et al. Driving behavior models enabling the simulation of advanced driving assistance systems: Revisiting the action point paradigm. *Transportation Research Part C: Emerging Technologies*, 36:352–366, 2013. DOI: 10.1016/j.trc.2013.09.009. 12

[69] Bing, Z., et al. Unsupervised clustering of driving styles based on KL divergence, *Automotive Engineering*, 40(11):1317–1323, 2018. 12

[70] Xu, L., Hu, J., Jiang, H., et al. Establishing style-oriented driver models by imitating human driving behaviors. *IEEE Transactions on Intelligent Transportation Systems*, 16(5):2522–2530, 2015. DOI: 10.1109/tits.2015.2409870. 12

[71] Bing, Z., Shude, Y., Jian, Z., and Weiwen, D. Personalized lane-change assistance system with driver behavior identification. *IEEE Transactions on Vehicle Technology*, 67(11):10293–10306, 2018. DOI: 10.1109/tvt.2018.2867541. 12

[72] Matthaeia, R., et al. *Autonomous Driving: Technical, Legal and Social Aspects*, Springer, Berlin, 2015. 12

[73] Bonnefon, J. F., Shariff, A., and Rahwan, I. The social dilemma of autonomous vehicles, *Science*, 352(6293):1573–1576, 2016. DOI: 10.1126/science.aaf2654. 12

[74] Abbink, D. A., Mulder, M., and Boer, E. R. Haptic shared control: Smoothly shifting control authority. *Cognitive Technology and Work*, 14:19–28, 2012. DOI: 10.1007/s10111-011-0192-5. 12

[75] Mulder, M., Abbink, D. A., and Boer, E. R. Sharing control with haptics: Seamless driver support from manual to automatic control. *Human Factors*, 54(5):786–98, 2012. DOI: 10.1177/0018720812443984. 12

[76] Saito, T., Wada, T., and Sonoda, K. Control authority transfer method for automated-to-manual driving via a shared authority mode. *IEEE Transactions on Intelligent Vehicles*, 3(2):198–207, 2018. DOI: 10.1109/tiv.2018.2804167. 12

[77] Nguyen, A.-T., Sentouh, C., and Popieul, J.-C. Sensor reduction for driver-automation shared steering control via an adaptive authority allocation strategy. *IEEE/ASME Transactions on Mechatronics*, 23(1):5–16, 2018. DOI: 10.1109/tmech.2017.2698216. 13

[78] Li, M., Cao, H., Song, X., Huang, Y., Wang, J., and Huang, Z. Shared control driver assistance system based on driving intention and situation assessment. *IEEE Transactions on Industrial Informatics*, 14(11):4982–4994, 2018. DOI: 10.1109/tii.2018.2865105. 13

[79] Erlien, S. M., Fujita, S., and Gerdes, J. C. Shared steering control using safe envelopes for obstacle avoidance and vehicle stability. *IEEE Transactions on Intelligent Transportation Systems*, 17(2):441–451, 2016. DOI: 10.1109/tits.2015.2453404. 13

[80] Tan, D., Chen, W., Wang, H., and Gao, Z. Shared control for lane departure prevention based on the safe envelope of steering wheel angle. *Control Engineering Practice*, 64:15–26, 2017. DOI: 10.1016/j.conengprac.2017.04.010. 13

[81] Nguyen, A.-T., Sentouh, C., and Popieul, J.-C. Driver-automation cooperative approach for shared steering control under multiple system constraints: Design and experiments. *IEEE Transactions on Industrial Electronics*, 64(5):3819–3830, 2017. DOI: 10.1109/tie.2016.2645146. 13

[82] Sentouh, C., Nguyen, A.-T., Benloucif, M. A., and Popieul, J.-C. Driver-automation cooperation oriented approach for shared control of lane keeping assist systems. *IEEE Transactions on Control System Technology*, 27(5):1962–1978, 2019. DOI: 10.1109/tcst.2018.2842211. 13

[83] Na, X. and Cole, D. J. Game-theoretic modeling of the steering interaction between a human driver and a vehicle collision avoidance controller. *IEEE Transactions on Human-Machine Systems*, 45(1):25–38, 2015. DOI: 10.1109/thms.2014.2363124. 13

[84] Na, X. and Cole, D. J. Application of open-loop stackelberg equilibrium to modeling a driver's interaction with vehicle active steering control in obstacle avoidance. *IEEE Transactions on Human-Machine Systems*, 47(5):673–685, 2017. DOI: 10.1109/thms.2017.2700541. 13

[85] Flad, M., Frohlich, L., and Hohmann, S. Cooperative shared control driver assistance systems based on motion primitives and differential games. *IEEE Transactions on Human-Machine Systems*, 47(5):711–722, 2017. DOI: 10.1109/thms.2017.2700435. 13

[86] Ji, X., Yang, K., Na, X., Lv, C., and Liu, Y. Shared steering torque control for lane change assistance: A stochastic game-theoretic approach. *IEEE Transactions on Industrial Electronics*, 66(4):3093–3105, 2019. DOI: 10.1109/tie.2018.2844784. 13

[87] Daskalakis, C., Goldberg, P. W., and Papadimitriou, C. H. The complexity of computing a nash equilibrium. *SIAM Journal on Computing*, 39(1):195–259, 2009. DOI: 10.1137/070699652. 20

[88] Simaan, M. and Cruz, J. B. On the stackelberg strategy in nonzero-sum games. *Journal of Optimization Theory and Applications*, 11(5):533–555, 1973. DOI: 10.1007/bf00935665. 20

[89] van Hasselt, H., Guez, A., and Silver, D. Deep reinforcement learning with double Q-learning. *National Conference on Artificial Intelligence*, pages 2094–2100, 2016. 22

[90] Schaul, T., Quan, J., Antonoglou, I., and Silver, D. Prioritized experience replay. *ArXiv: Learning*, 2015. 22

[91] Wang, Z., Schaul, T., Hessel, M., van Hasselt, H., Lanctot, M., and de Freitas, N. Dueling network architectures for deep reinforcement learning. *International Conference on Machine Learning*, pages 1995–2003, 2016. 22

[92] Hessel, M., et al. Rainbow: Combining improvements in deep reinforcement learning. *National Conference on Artificial Intelligence*, pages 3215–3222, 2018. 22

[93] Min, K., Kim, H., and Huh, K. Deep distributional reinforcement learning based high-level driving policy determination. *IEEE Transactions on Intelligence Vehicles*, 4(3):416–424, 2019. DOI: 10.1109/tiv.2019.2919467. 24

[94] Min, K., Kim, H., and Huh, K. Deep Q learning based high level driving policy determination. *IEEE Intelligent Vehicles Symposium (IV)*, 2018. DOI: 10.1109/ivs.2018.8500645. 24

[95] Zhang, M., Li, N., Girard, A., and Kolmanovsky, I. A finite state machine based automated driving controller and its stochastic optimization. *ASME Dynamic Systems and Control Conference*, 2017. DOI: 10.1115/dscc2017-5209. 26

[96] Peng, J., Guo, Y., Fu, R., Yuan, W., and Wang, C. Multi-parameter prediction of driverslane-changing behaviour with neural network model. *Applied Ergonomics*, 50:207–217, 2015. DOI: 10.1016/j.apergo.2015.03.017. 89

[97] Chapelle, O., Sindhwani, V., and Keerthi, S. S. Optimization techniques for semi-supervised support vector machines. *Journal of Machine Learning Research*, 9:203–233, 2008. 95

[98] Lei Wu, M.-L. Z. Multi-label classification with unlabeled data: An inductive approach. *ACML*, pages 197–212, 2013. 96

[99] Li, M., et al. Shared control with a novel dynamic authority allocation strategy based on game theory and driving safety field. *Mechanical Systems and Signal Processing*, 124:199–216, 2019. DOI: 10.1016/j.ymssp.2019.01.040. 97

[100] Wang, J., Wu, J., Zheng, X., Ni, D., and Li, K. Driving safety field theory modeling and its application in pre-collision warning system. *Transportation Research Part C: Emerging Technologies*, 72:306–324, 2016. DOI: 10.1016/j.trc.2016.10.003. 97

[101] Na, X. and Cole, D. J. Linear quadratic game and non-cooperative predictive methods for potential application to modelling driver—AFS interactive steering control. *Vehicle Sytems Dynamics*, 51(2):165–198, 2013. DOI: 10.1080/00423114.2012.715653. 101

[102] Rawlings, J. B. and Mayne, D. Q. *Model Predictive Control: Theory and Design*, Nob Hill Pub. Madison, Wisconsin, 2009. 103

[103] Li, G., Yang, Y., and Qu, X. Deep learning approaches on pedestrian detection in hazy weather. *IEEE Transactions on Industrial Electronics*, 2019. DOI: 10.1109/tie.2019.2945295. 106

[104] Cao, H., Song, X., Zhao, S., Bao, S., and Huang, Z. An optimal model-based trajectory following architecture synthesising the lateral adaptive preview strategy and longitudinal velocity planning for highly automated vehicle. *Vehicle Sytems Dynamics*, 55(8):1143–1188, 2017. DOI: 10.1080/00423114.2017.1305114. 108

[105] Mamdani, E. H. and Assilian, S. An experiment in linguistic synthesis with a fuzzy logic controller. *International Journal of Man-Machine Studies*, 7(1):1–13, 1975. DOI: 10.1016/b978-1-4832-1450-4.50032-8. 112

Authors' Biographies

HAOTIAN CAO

Haotian Cao is currently a Postdoctoral Fellow at the College of Mechanical and Vehicle Engineering, Hunan University, Changsha, China. He received a B.E. in vehicle engineering and a Ph.D. in mechanical engineering from the College of Mechanical and Vehicle Engineering, Hunan University, Changsha, China, in 2011 and 2018, respectively. He was a visiting scholar at Human Factors group, the University of Michigan Transportation Research Institute (UMTRI) from 2016 to 2017. He is a committee member of the Chinese Association of Automation Parallel Intelligence (2018-2022), and a referee of multiple international journals and conferences. He is also a Principal Investigator (PI) of a project funded by Natural Science Foundation of China (NFSC). His interests include trajectory planning and following control for autonomous vehicles, technology related to vehicle dynamical systems, driver behavior modeling, and naturalistic driving data analysis.

MINGJUN LI

Mingjun Li received a B.E in vehicle engineering from Hunan University, Changsha, China, in 2016, where he has been working toward a Ph.D. in mechanical engineering with the College of Mechanical and Vehicle Engineering since 2017. He is also a visiting Ph.D. student with the Waterloo Cognitive Autonomous Driving (CogDrive) Lab at University of Waterloo, Canada from September 2019. His research interests include the shared control strategy, vehicle dynamics and control, driver-assistance systems, and human-driver behaviors analysis for intelligent vehicles.

SONG ZHAO

Song Zhao received a B.E. in vehicle engineering from Hunan University, Changsha, China in 2016. He then studied at the University of Michigan-Ann Arbor and received an MEng. in Global Automotive Manufacturing Engineering. After working in the industry for three years, he is now pursuing his Ph.D. in the Waterloo Cognitive Autonomous Driving (CogDrive) lab at Univesity of Waterloo, Canada. His research interests include vehicle dynamic and control, behavior and motion planning, and advance driver-assistance systems for intelligent vehicles.

XIAOLIN SONG

Xiaolin Song received her B.E., M.E., and Ph.D. at the College of Mechanical and Vehicle Engineering, Hunan University in 1988, 1991, and 2007, respectively. From 2008 to the present, she has been a professor and a Ph.D. supervisior at Hunan University. She was an advanced visiting scholar of the University of Michigan (Ann Arbor), the University of Waterloo, and the University of Texas at Austin. She is a Vice Chairman of Rules Committee of Formula Student China, as well as an Academic Committee Member of the College of Mechanical and Vehicle Engineering, Hunan University. She has been an independent PI and Co-PI for over five projects of the NSFC, and over ten other provincial and ministerial projects. Her research interests include vehicle active safety, vehicle dynamics control, driver modeling, and human factors in driving safety.

Printed in the United States
by Baker & Taylor Publisher Services